CATALYST HANDBOOK

CATALYST HANDBOOK

WITH SPECIAL REFERENCE TO
UNIT PROCESSES IN AMMONIA
AND HYDROGEN MANUFACTURE

1970

SPRINGER-VERLAG NEW YORK INC.

WOLFE SCIENTIFIC BOOKS/ LONDON – ENGLAND

SOLE DISTRIBUTORS FOR THE
UNITED STATES, CANADA AND MEXICO
SPRINGER-VERLAG NEW YORK INC./ NEW YORK

Library of Congress Catalog Card Number: 70-121128
Title Number: 8-80017
SBN 72340164 0

© Imperial Chemical Industries Limited 1970

No part of this book may be reproduced in any form
by print, photoprint, microfilm, or any other means,
without permission from the publisher

Printed in Great Britain by Billing & Sons Ltd,
Guildford and London

Contents

PREFACE — vii

1 CATALYTIC ACTIVITY OF SOLIDS — 1
  By D. A. Dowden, B.Sc., former visiting Professor in Department of Chemical Engineering and Chemical Technology, Imperial College of Science and Technology, London.
  The activity of solids, 5; The specific area, 6; The intrinsic efficacy, 7.

2 THE STRUCTURAL ENGINEERING OF CATALYSTS — 20
  By Professor S. P. S. Andrew, M.A., B.Sc.(Eng.), M.I.Mech.E., A.M.I.Chem.E.
  High specific surface, 21; The influence of internal and external mass transfer in catalyst design, 22; The structural properties of components, 23; Geometric and strength factors, 24; Poisoning and structural collapse, 28; Other constructional processes, 28.

3 CATALYST TESTING — 31
  By J. S. Campbell, B.Sc., Ph.D.
  Rate-limiting steps in gas–solid reactions, 33; Test reactor types, 38; Strength and attrition testing, 44.

4 DESULPHURISATION — 46
  By J. J. Phillipson, B.Sc., Ph.D.
  Feedstock for steam reforming, 46; Methods of desulphurisation, 47; ICI catalysts, 48; Some principles of desulphurisation, 55; Deactivation of cobalt molybdate, 61; Future developments, 63.

5 HYDROCARBON-REFORMING CATALYSTS — 64
  By G. W. Bridger, B.Sc., A.R.I.C., and G. C. Chinchen, B.A. Cantab.
  Thermodynamics of reforming, 65; Reforming catalysts, 77; Catalyst poisons, 83; Catalyst performance, 87; Kinetics and mechanism, 90; Calculation of equilibrium compositions, 94.

6 REMOVAL OF CARBON MONOXIDE — 97
  By J. S. Campbell, B.Sc., Ph.D., P. Craven, B.Sc.(Eng.), A.C.G.I., M.Sc., D.I.C., and P. W. Young, B.Sc.
  The water-gas shift reaction, 97; High-temperature shift catalysts, 98; Reactions with HT shift catalysts, 100; Low-temperature shift

catalysts, 109; Reactions on LT shift catalysts, 113; Development of new LT shift catalysts, 117; Methanation, 117; Methanation catalysts, 119; Kinetics of methanation, 121.

7   AMMONIA SYNTHESIS CATALYSTS   126
    By G. W. Bridger, B.Sc., A.R.I.C., and C. B. Snowdon, M.A., Ph.D. Cantab.
    Catalyst formulation, 131; Catalyst poisons, 138; Catalyst reduction, 139; Prereduced catalyst, 140; Kinetics of ammonia synthesis, 141.

8   COMPUTER PROGRAMS FOR CONVERTER   147
    CALCULATIONS
    By W. Spendley, B.Sc., F.S.S.
    Optimisation, 148; Requirements of an effective program, 150; Shift converter programs, 151; Ammonia converter programs, 155; Other programs, 160.

9   HANDLING AND USING CATALYSTS ON THE PLANT   161
    By D. R. Goodman, B.A. Oxon.
    Catalyst reduction, 168; Regeneration of catalysts and blanketing procedures, 174; Catalyst discharge, 176; Safety precautions, 181.

GENERAL REFERENCES ON CATALYSIS   182

REFERENCES   183

APPENDICES   185

INDEX   223

Preface

THIS book deals with heterogeneous catalysis by solids, and its material is drawn almost entirely from the science and technology of the unit processes used in ammonia and hydrogen manufacture.

In the last few years a rapidly growing demand for fertilizers, and the success of the ICI steam-naphtha reforming process, have resulted in the construction of hundreds of new chemical plants. These have necessarily been staffed largely by technologists with no previous experience in the management of catalysts and catalytic processes.

At the same time, the size of single-stream plants has risen dramatically, principally because of advances in catalysis chemistry. The designers and operators of such plants have been content to proceed on the assumption that a catalyst is a reagent delivered in a drum bearing a label stating how the contents should be used, and how (given certain conditions) it will perform. The consumers have been content to regard the chemistry of a catalyst as a mystery best left to the experts employed by the company making the stuff.

This was a reasonable working attitude when the practice of catalysis was still a matter of art and empiricism. But those days are past. We now know enough about the shape and the function of these chemical entrepreneurs to make possible a rational approach to their design and use, and one purpose of this book is to give every catalyst user a handy source of practical information and an intelligent appreciation of the nature of the tools he employs.

The editors would like to thank the Board of Imperial Chemical Industries Limited Agricultural Division, for permission to publish the original work described in this book, most of which was done in the laboratories of the division, and Technical Co-ordinator P. W. Young, B.Sc., for his work on the text.

CHAPTER 1

Catalytic activity of solids

CHEMISTS have known for over 150 years that many reactions are quickened in the presence of certain substances called catalysts. Good catalysts can induce the rapid transformation of a great many reactant molecules while themselves apparently suffering little alteration in character or quantity. If the catalyst exists in the same phase as the reactants it is said to be homogeneous; otherwise it is heterogeneous. In practice, this latter term is usually reserved for systems involving solid catalysts—the principal subjects of this volume.

In the chemical industry, catalysts are used in order to bring some chosen reaction as close as possible to a selected equilibrium point in the shortest possible time. Fig. 1 sets out the process involved.

The reaction
$$v_1 R_1 + v_2 R_2 + \ldots v_j R_j \rightleftharpoons v_{j+1} P_1 + \ldots v_c P_c \qquad \text{(i)}$$
(The R's and P's are reactants and products respectively and the v's are their molar coefficients) proceeding at a temperature $T(°K)$ will have an appreciable equilibrium constant (K), where
$$K = \frac{[P_1]^{v^{j+1}}[P_2]^{v^{j+2}} \ldots [P_c]^{v^c}}{[R_1]^{v^1}[R_2]^{v^2} \ldots [R_j]^{v^j}}, \qquad \text{(ii)}$$
only if the free energy change (ΔF_T) is small or negative:
$$\Delta F_T = -RT \ln K. \qquad \text{(iii)}$$
The free energy change is given by
$$\Delta F_T = \Delta H_T - T\Delta S_T, \qquad \text{(iv)}$$
where ΔH_T and ΔS_T are respectively the changes in enthalpy and entropy and R is the gas constant.
† square brackets indicate concentrations.

FIG. 1. General scheme for a reaction proceeding to equilibrium.

The problem presents itself as a need to devise, improve, or cheapen a catalytic process, and because of all the factors present the catalyst usually has the greatest influence upon space-time yields, it is of prime importance. In both batch and continuous processes the reactants involved are not at equilibrium during a large fraction of the time at which they are in contact with the catalyst. Therefore classical thermodynamics contribute little to our understanding of how catalysts work. Two trivial conclusions can, however, be drawn. The first is that catalysts cannot cause systems to move away from equilibrium. The second is that for a reaction at equilibrium a catalyst must increase the rates of the forward and reverse reactions in the same ratio. Hence the rough rule, in general accord with observation, that even away from equilibrium a reaction and its reverse are both accelerated by the same catalysts. This rule has no quantitative value in practice and applies only if the catalyst remains essentially unchanged over a range of non-equilibrium conditions.

Industrial catalytic processes are usually complex in that the reactants may be converted to more than one product by a series of parallel and consecutive reactions. The thermodynamics of each step and of the overall process can be given by equations like (i) to (iv) in fig. 1, and so it is convenient to classify steps or processes as:

(a) in practice reversible or irreversible under the operating conditions, i.e. $-\Delta F_T$ small or large respectively;
(b) giving the desired substances either as end-products or intermediates, i.e. according to whether or not the desired products are formed by the reaction with the largest free energy change;
(c) endothermal (positive ΔH_T) or exothermal (negative ΔH_T) and so having larger K's at higher and lower temperatures respectively; or
(d) composed of coupled reactions. In these an expendable product of a desired reaction which is normally endothermal (possessing a suitable $-\Delta F_T$ only at high temperatures) is alone combined in the succeeding step with an auxiliary reactant chosen to make the process exothermal overall and thus feasible at lower temperatures. The succession of reaction steps in complex processes is not necessarily determined by the ΔH's; in coupled reactions it is.

The rates (r) of catalysed reactions cannot be calculated, or even estimated, from the fundamental atomic constants but must be determined by experiment. They are found to depend upon the concentration of reactants and products in more or less complicated forms.

$$r = k\,\mathrm{f}([R_j], [P_c]), \qquad \text{(v)}$$

where k is the rate constant.

The function is often a product of powers of the concentrations or sums of the products, and the powers may be positive or negative and integral or fractional (usually $\not> 2$). The temperature coefficient of the rate constant is closely approximated, at least over small ranges of temperature, by the relation $k = a\exp(-b/T)$ where a and b are constants. Although examples are known of reactions which decrease in rate above some temperature, a rise in temperature usually greatly increases rates, and the temperature of reaction is a most important variable over which it is essential to have control.

The empirical relation is usually replaced by the Arrhenius equation

$$k = A\mathrm{e}^{-E/RT} \qquad \text{(vi)}$$

in which A is the frequency factor and E the activation energy per mole. However, in heterogeneous catalysis the constants of this equation can seldom be directly associated with the activated complex of a rate-controlling step;

they also contain terms which depend upon changes in the surface concentration of the adsorbates and in the solid phases of the catalyst.

At the inlet to a flow reactor, or at zero time for a batch reactor, the reactants are at their initial concentrations, whereas towards the exit of the flow reactor, or the termination of the batch reaction, the concentrations are lower, being possibly close to or at their equilibrium values. The rate (v) of reaction at any time is the difference between the forward (v_1) and reverse (v_{-1}) rates, i.e.

$$v = v_1 - v_{-1}$$
$$= k_1 f_1([R_j], [P_c]) - k_{-1} f_{-1}([R_j], [P_c])$$
$$= k_1 \{f_1([R_j], [P_c]) - f_{-1}([R_j], [P_c]) k_{-1}/k_1\}$$

Initially therefore $v_0 = v_1 = k_1 f_1([R_j], [P_c])$ but with increasing conversion

$$v = v_1 - v_{-1} = k_1 \{f_1([R_j], [P_c]) - ([R_j], [P_c]) f_{-1}/K\}$$

The rate 'constants' k_1 and k_{-1} may depend upon the concentrations ($[R_j], [P_c]$) so that the best catalyst for small conversions may not be the best for large conversions. As reactors are seldom isothermal the real situation is still more complicated. To meet this situation the use of reactors with beds of catalysts of graded characteristics has been suggested.

The rate of a catalysed reaction can therefore be changed by varying the temperature, the catalyst, or the concentrations of reactants and products. Although these variables are not independent, the responses are especially sensitive to the catalyst variations, because each solid introduces its unique kinetics and rate constants.

The reactions of nitrogen and hydrogen can be written as a virtual mechanism:

$$N\equiv N + H-H \rightleftharpoons HN=NH \underset{2H_2}{\overset{H_2}{\rightleftharpoons}} H_2N-NH_2 \overset{H_2}{\rightleftharpoons} 2NH_3 \overset{H_2}{\rightleftharpoons} 2NH_4 \overset{H_2}{\rightleftharpoons} 2NH_5$$

(with $3H_2$ overall)

but thermodynamics and the principles of chemistry reduce these to $\frac{1}{2}N_2 + \frac{3}{2}H_2 = NH_3$. Ammonia synthesis is an exothermic, reversible reaction to be brought as close as possible to equilibrium for the most stable product. Appropriate catalysts must have the highest activity for the hydrogenation of nitrogen at as low a temperature as possible in order to take advantage of the higher equilibrium constants.

The saturation of acetylene can be represented by a similar series of steps:

$$CH\equiv CH + H-H \rightleftharpoons H_2C=CH_2 \overset{H_2}{\rightleftharpoons} H_3C-CH_3 \overset{H_2}{\rightleftharpoons} 2CH_4$$

(with $3H_2$ overall, $2H_2$)

but now all the intermediates are sufficiently stable to be isolated in high yield. The hydrogenation of acetylene to ethylene is an exothermic, reversible reaction, but the olefine is not as stable as the paraffins, so the catalyst must be active for the first step but inactive for the subsequent steps, i.e. the catalyst must be 'selective'.

The union of carbon monoxide and hydrogen displays the same broad features:

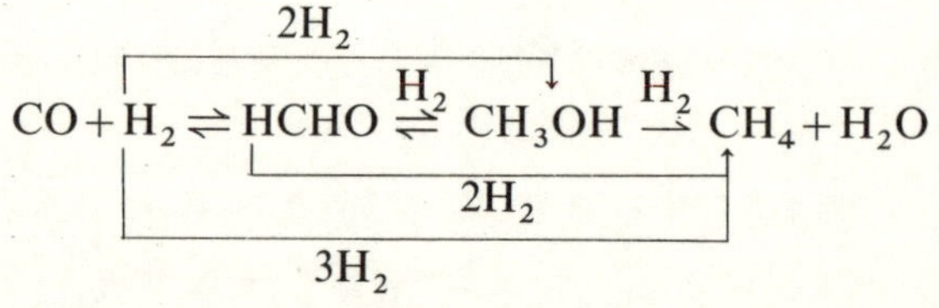
2H$_2$
H$_2$
H$_2$
CO + H$_2$ ⇌ HCHO ⇌ CH$_3$OH → CH$_4$ + H$_2$O
2H$_2$
3H$_2$

Formaldehyde is a virtual intermediate because the data show that its equilibrium partial pressure must be very small at reasonable temperatures. The methanation reaction gives the very stable molecules CH_4 and H_2O in an exothermic reaction for which again an active catalyst without selectivity is required. The production of methanol is possible only over a selective catalyst and because ΔH is negative it should be active at the lowest possible temperatures.

The foregoing illustrate the necessity for catalyst selectivity in the control of consecutive reactions. The decomposition of isopropanol demonstrates the control of parallel endothermic reactions:

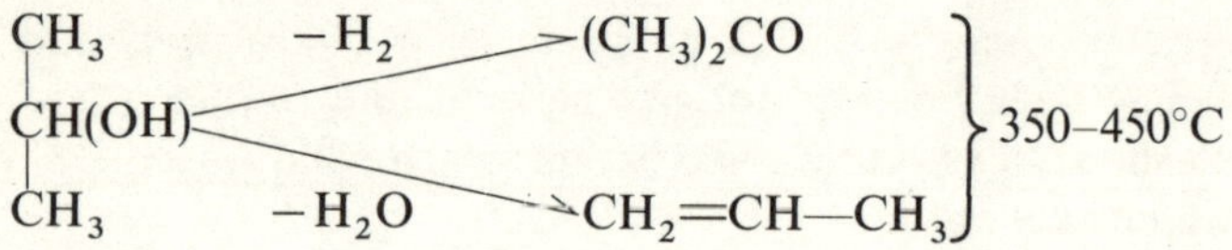
CH$_3$
CH(OH)
CH$_3$
−H$_2$
(CH$_3$)$_2$CO
−H$_2$O
CH$_2$=CH—CH$_3$
350–450°C

Catalysts can be chosen to direct this reaction to give almost 100 per cent dehydrogenation or dehydration but because the equilibrium constants are larger at higher temperatures other parasitic reactions (not shown) may intrude under working conditions.

Oxidations are typical of irreversible, highly exothermic reactions which may be conducted to yield either the most stable products (such as the conversion of ammonia to nitrogen and water) or intermediate products—selective oxidation (as with the conversion of ammonia to oxides of nitrogen, of methane to formaldehyde, and of propylene to acrolein).

The dehydrogenation of butene to butadiene

$$CH_3{-}CH_2{-}CH{=}CH_2 \rightleftharpoons CH_2{=}CH{-}CH{=}CH_2 + H_2$$

requires a temperature near 500°C to attain a suitable equilibrium constant. Many solids have appropriate activities for this reaction but the presence of large concentrations of unsaturated hydrocarbons can lead, under these conditions, to undesirable parasitic processes which can be minimised only if the catalyst is to some degree selective. By coupling with the oxidation of hydrogen, the reaction becomes an oxidative dehydrogenation

$$CH_3{-}CH_2{-}CH{=}CH_2 + \tfrac{1}{2}O_2 \rightarrow CH_2{=}CH{-}CH{=}CH_2 + H_2O$$

The equilibrium constant is now large at lower temperatures, but the catalyst must be selective to avoid burning.

It is evident that the selection of suitable catalysts depends upon the extent of knowledge and the correctness of suppositions concerning the nature of the reaction and of the activity of solids.

The activity of solids

The catalyst, as separate particles or agglomerates of particles, is immersed in a fluid medium in motion. Reactants and products diffuse in the gas or liquid phases bathing the boundaries of the solid, and also in the pore spaces of the aggregates. These diffusions may be rate-controlling but are not considered in detail here. The reactant, the intermediates, and the products combine loosely (physisorption) or more tightly (weak and strong chemisorption) with the surface of the solid. Solubilities and rates of diffusion in solids are small, so the reaction almost invariably takes place on the solid surface and involves solid–substrate interactions which stretch or dissociate the bonds of the reactants. Thus at least one of the reactants must be chemisorbed, but subsequent steps may occur through collisions between adsorbed species, or between these and molecules impinging directly from the fluid.

The overall rate can therefore be controlled by diffusion, adsorption, desorption, or by an interaction between surface complexes in a simple reaction, or at some intermediate step in a complex process.

The principal characteristics

The efficiency of a catalyst rests upon three components—activity, selectivity, and life—which may not function independently.

The activity is the extent to which the catalyst influences the rate of change of the degree of advancement of the reaction (as assessed by the disappearance of reactants—the conversion), per unit weight or per unit volume of catalyst, under specified conditions. It is best expressed by the kinetic equation (v), but is often available only as a rate (r), a rate relative to some standard or a qualitative and subjective estimate. The activity per unit volume is of practical importance because process economics can depend critically upon the cost of packed reactor space: the bulk density of the catalyst must always be as small as possible, consistent with other requirements.

The life of the catalyst is the period during which the catalyst produces the required product at a space-time yield in excess of or equal to that designated. The activity of most catalysts decreases sharply at first and then declines much more slowly with time. The selectivity may worsen or improve. The life of a catalyst terminates because of loss of mechanical strength or because of unacceptable changes in activity and selectivity.

The catalytic efficacy of a homogeneous solid phase, crystalline or amorphous, is influenced by the four factors (a)–(d) listed below.

(a) The exposed area in contact with the fluid.
   The area of the isolated phase is determined by the physisorption of an inert gas (N_2, A, Kr) and when expressed per unit weight of catalyst is called the specific area.
   Activity and selectivity scaled to unit area are also labelled 'specific'.
(b) The intrinsic chemical characteristics of the surface of the solid.

The species forming the surface may be atoms or ions and their chemical properties must depend upon their electronic structure and arrangement, as, for example, their electronic configuration, co-ordination number and local symmetry.

(c) The topography of the surface.

Because of the dependence of activity upon geometry and electronic structure, the faces, edges, and corners of crystals must possess different activities.

(d) The occurrence of lattice defects.

The activities to be associated with defects such as vacancies, interstitials, dislocations, grain boundaries, etc., are currently undecided but they must differ from solid to solid and with the type of reaction catalysed.

Obviously, the four factors are interdependent, and only in special circumstances can one be varied without comparable changes in the rest.

In multicomponent catalysts each homogeneous phase will exercise its characteristic activity. The phase boundaries may have different properties which, it is simplest to assume, are associated with the local formation of solid solutions or compounds by solid state reactions. Chemisorption and X-ray methods can sometimes be used to determine the surface areas of the several phases in composities.

The specific area

The specific area (S) of the solid phases of the catalyst must be adjusted to suit the requirements of the process. Usually the areas are made and maintained large by procedures which produce either small particles or porous bodies in more or less stable states. The particles may be used as such but are most often formed into larger aggregates ($\frac{1}{16}$–1 in) with some small reduction in area. The relationship between S and particle diameter (d) is given for a simple cubic packing of regular, non-porous smooth spheres (or of slightly separated cubes, of edge d) by the equation

$$S = 6/\rho d$$

where ρ is the density of the particles, usually taken to be that of the solid in bulk. The model of spheres, in which the pore diameters must be smaller than d, shows that small pores coexist with large areas and introduce the possibility of a pore diffusion limitation on the reaction rate. The cube model illustrates that larger areas involve longer lengths (L) of edges

$$L = \frac{12}{\rho d^2} = \frac{2S}{d}$$

and greater numbers (N) of corners

$$N = \frac{8}{\rho d^3} = \frac{4S}{3d^2}$$

and thus the introduction of more regions of different activity.

Real catalysts are composed of particles of a range of sizes, and it is well known that if movement of lattice components (leptons) is possible, then changes must ensue which minimise the surface energy. In isolated particles of low vapour pressure this can occur only by alterations of shape (smoothing) to expose faces of low energy but in ensembles, in addition, large particles grow at the expense of small. Thus S, L, and N all decrease, and the smaller the particles, the more rapidly does this happen. These processes, which have come to be known collectively as 'sintering', must be hindered, as they are always harmful to specific activity, if not to selectivity.

The movement of leptons can occur within the solid, in a thin peripheral layer, on the surface, in a thin adsorbed layer, or through the fluid phase. Energy is required to move leptons from stable or metastable positions and hence their rates of migration increase exponentially with temperature.

Temperature has therefore a marked but at present incalculable effect upon S, N, and L. In estimating these effects, recourse is had to the old empirical rules which state that lattices begin to be appreciably mobile in the bulk at $0.5\ T_m$ (Tammann) and in the surface at $0.3\ T_m$ (Hüttig) where T_m is the melting point (°K) of the solid. The rules are rough and appear to work best for simple solids without phase changes and of low vapour pressure.

Sintering can be restricted to smoothing by dispersing the particles of the active phase on the surface of another inert refractory solid of high area (support action), or by separating them with refractory spacing blocks (stabilisation). But smoothing itself cannot be controlled in practice. Migration of a catalyst component is facilitated if it is soluble in the fluid or can form, in situ, a soluble phase. For instance, some transitional metals can be transported as volatile carbonyls, halides and oxides, while many 'insoluble' oxides and salts have sufficient solubility in liquids (especially multilayers of water), or stability as gaseous hydrates, to accelerate sintering. This phenomenon is especially insidious because it can occur at small partial pressures of adventitious impurities, quite sufficient to effect crystal growth and to move material up or down temperature gradients, but not to cause carry over from the reactor (ref. 1).

Finally, surface area can be obscured by debris (dust, rust) or encapsulated by such products of parasitic reactions as liquid polymers and solid 'coke'. If thereby the pore-size distribution is changed and the reactions become diffusion limited, then the selectivity as well as the activity may be impaired.

The intrinsic efficacy

Catalysts under reaction conditions may not contain the same solid phases as were present when the reactor was charged. Because the activity is controlled by the phases existing in situ, these should be known, if efficacy is to be predicted and adjusted. Frequently the details can only be guessed.

The intrinsic efficacy might be thought to vary with particle size, yet current researches suggest that for many metals the efficacy is an invariant. However, this can hardly be true for all solids, even when they are effecting their typical catalyses. Nevertheless, the intrinsic efficacy is sufficiently characteristic to enable a useful classification of solids to be made in terms of their specificity and activity.

Classification of reactions

Fig. 2 provides a short classification of reactions for which catalysts are employed. The examples chosen are from among the commoner types which are of industrial importance as overall reactions, or as virtual intermediate steps.

Redox reactions
Hydrogenation–dehydrogenation, hydrogen exchange (disproportionation), hydrogenolysis.

Oxidative dehydrogenation, oxygen (sulphur) insertion, combustion.

Substitution
Halogenation, amination, hydrolysis.

Addition——elimination
Hydration–dehydration, hydrohalogenation, hydrosulphidation, amination, carbonylation, carboxylation, alkylation, condensation–depolymerisation, polymerisation; also their reverse reactions, e.g. dealkylation ('cracking').

Cyclisation
Aromatisation (dehydrogenation, dehydration, dehydrosulphidation, etc.), ring closure to anhydrides, cyclic ethers, sulphides, imines, etc.

Molecular rearrangements
Multiple-bond shift, skeletal isomerisation, ring contraction or enlargement.

FIG. 2. Some industrially important reactions for which catalysts are employed. The stoichiometry describes the reaction to be accelerated or inhibited.

The rules of thumb of chemistry apply also to these reactions over solids. For example unsaturated compounds are more reactive than saturated compounds, aromatic substances are relatively stable, as are other intermediates of high resonance energy, polar additions to olefines obey Markownikow's rule, carbonium ions react according to the accepted patterns, and so on. Moreover current notions imply a much greater similarity between heterogeneous and homogeneous mechanisms than has hitherto been taken into account. Redox reactions are effected by catalysts which are themselves redox systems. Substitution, addition-elimination and molecular rearrangement are susceptible to catalysis by acids and bases.

The patterns of efficacy

It is generally impossible to predict the absolute efficacy of catalysts for specific reactions, but comparative efficacies can be shown to fall into regular patterns. For instance the hydrogenations of different multiple bonds over metals occur at very different rates but these are always much greater over group 8 metals than over group 1B metals.

Chemical properties together with early atomic theory led to the recognition of two kinds of chemical bond—the ionic and the covalent—and also the association of stability and inertness with the closed-shell electron configurations of the inert gases (ns^2, ns^2np^6, $(n-1)d^{10}ns^2np^6$, etc.). Later the metallic bond became better understood and a measure of stability was found to be associated also with half full and full sub-shells of electrons (e.g. nd^5, nd^{10}).

The ions of the elements of the short periods and of the ends of the long periods tend to have an invariant valency, and to adopt closed shell structures forming colourless diamagnetic compounds which are typical insulators (MgO, Al_2O_3, SiO_2). The transitional elements have variable valencies, form stable ions with unfilled d-shells, and may give coloured, magnetic, semi-conducting compounds wherein the cations have electrons with unpaired spins. The ions with half full or full sub-shells (d^5, Mn^{2+}; d^{10}, Zn^{2+}) may be said to fall between these two extremes in properties.

Thus breaking a covalent bond to give two neutral species must always produce two radicals, each with a free valency and possessing the activity of free radicals. Breaking an ionic bond may give either two closed shell ions having only electrostatic polarising power ($MgO = Mg^{2+} + O^{2-}$) or two ions one of which (usually the cation) has also unpaired electron spin, and therefore additional, radical-like properties (e.g. $NiO = Ni^{2+} + O^{2-}$). Solid surfaces prepared by cleavage in vacuum present arrays of broken bonds which can react with molecules from the fluid phase (chemisorption) to give various surface complexes, the thermodynamic and kinetic properties of which determine the course of the catalysis.

Clearly the catalytic efficacy of a solid depends upon its constituent leptons.

Early workers correlated high catalytic activity with variable valency, colour, magnetic properties, and so on. Still more recently the kind of electrical conductivity of the solid formed the dominant interrelation. These relations are best referred to electronic structure against the background of the periodic table, although this gives a superficial classification which is no substitute for a detailed knowledge of the chemistry and the physics of the solid phases involved.

Fig. 3 shows those solid elements which have unstable oxides or oxides reducible at moderate temperatures.

Conductors			Insulators
Metals		Semiconductors	Covalent
Transitional*	Non-transitional	Non-transitional	Non-transitional
		C (graphite)	C (diamond)
Fe Co Ni	Cu Zn S As Se	Ge	
Mo Ru Rh Pd†	Ag† Cd Sn Sb Te	Sn	
W Re Os Ir† Pt† and alloys	Au† Hg Pb Bi		

*i.e. having unfilled d orbitals.   † form only unstable oxides.

FIG. 3. Solid elements which have unstable oxides, or oxides reducible at moderate temperatures.

Fig. 4 lists some of the oxides which are representative of those found in catalysts. A similar table could be composed for sulphides but would present still greater difficulties in the representation of the phases. The tabulation in fig. 4 is not complete, as many of the intermediate oxides likely to exist in catalysts are non-stoichiometric (e.g. $V_{12}O_{26}$, Cr_2O_5, MoO_2, Bi_2O_3, etc.) and little is known about their semi-conducting properties. The oxides of elements stable in their highest group valency state (in moderate partial pressure of oxygen) are usually n-type semi-conductors (e.g. TiO_2, V_2O_5 and ZnO), while those stable in a lower valency state are p-type (e.g. Cr_2O_3, MnO, NiO). Systems in which the elements can have a series of valencies (e.g. Pb–O) often form a series of oxide phases each with a range of homogeneity extending on both sides of the stoichiometric composition so that the oxide can be either n-type or p-type depending upon the partial pressure of oxygen. Alternatively, ions of valency differing only by unity can coexist on the same set of lattice sites. The conductivity is then intrinsic because electrons can hop along a chain of cations as in Fe_3O_4.

The halides do not form non-stoichiometric salts, and their classification is not based upon conductivity but upon fixed or variable valency in halides of non-transitional and transitional elements (compare, for example, $MgCl_2$, $CuCl_2$, $HgCl_2$).

It is relevant to point out that at high temperatures, especially with 'active', finely divided solids, even the most stable insulators of section A in fig. 4 can become electronic conductors due to loss of lattice oxygen. The insulators range from the strongly basic through the amphoteric to the strongly acidic. However, it must not be overlooked that even the semi-conductors have basic or acidic properties—the oxides of lower valency are more basic than those of higher valency whether formed from different elements (e.g. ZnO v. WO_3), or from the same element (e.g. Cr_2O_3 v. CrO_3). Other trends, such as the stability of oxides, peroxides, peracids, ammines, and so on, must also be considered in catalyst design.

Ternary oxides, where formed between members of the same section, tend to retain *broadly* the conductivity-type of the binary oxides. Combinations between an insulator and a semi-conductor tend to behave like the semi-conductor, but the properties of compounds formed from semi-conductors of different types are not so predictable.

The efficacy of metals

Because industrial processes operate with gases containing water vapour, the number of metals remaining unoxidised in extended use, under reducing conditions, and at moderate temperatures, is confined to those listed in fig. 3, with the exception of Zn, and perhaps Mo and W. Under oxidising conditions far fewer can persist. Those that can are distinguished in the figure by the superscript [†].

(i) Chemisorption (ref. 2)

The chemisorptions of the simpler gases by the metals listed in fig. 3 all follow the same basic pattern. Without exception, the heats of chemisorption at small surface coverage, and the amounts adsorbed at saturation, increase on proceeding to the left in each long period. Also the more reactive gases are the more strongly and extensively held. Oxygen is strongly adsorbed by all

Insulators (electronic)		Semiconductors			
Non-transitional	Transitional				Non-transitional
A	B	C n-type	D p-type	E intrinsic	F n-type
BeO B$_2$O$_3$					
MgO Al$_2$O$_3$ SiO$_2$ P$_2$O$_5$					
CaO	Sc$_2$O$_3$	TiO$_2$ V$_2$O$_5$ Fe$_2$O$_3$	Cr$_2$O$_3$ MnO FeO CoO NiO Cu$_2$O	Fe$_3$O$_4$ Co$_3$O$_4$ CuO	ZnO GeO$_2$ As$_2$O$_5$
SrO	ZrO$_2$	Nb$_2$O$_5$ MoO$_3$			CdO SnO$_2$ Sb$_2$O$_5$
BaO	HfO$_2$	Ta$_2$O$_5$ WO$_3$			HgO PbO$_2$ Bi$_2$O$_5$
		UO$_3$			

FIG. 4. Solid binary oxides.

metals except gold. All other molecules are adsorbed weakly by the non-transitional metals, but much more strongly by the transitional metals. Inert molecules like N_2 and CO_2 are different only in that strong adsorption first appears further to the left in group 8, near iron and its congeners. It is known from the exchange of isotopes that the chemisorption of all molecules is accompanied by some degree of bond weakening or fission, so that metals have the prerequisites to function as catalysts for many reactions.

(ii) *Selectivity*

The reactions most typical of the selectivity of metals are the redox reactions, but they will also cause amination, decarbonylation, polymerisation, cyclisation and molecular rearrangements (see p. 8). Within this class of catalyses some metals are highly specific for certain reactions, as, for example, the partial saturation of triple bonds, or of one type of unsaturated bond in the presence of others, and analogous selective oxidations. A measure of specificity can usually be obtained by devices which diminish the activity of a given metal, or by choosing a less active metal, but, obviously, the ideal to be aimed for is high specificity associated with high activity. Thus, in the selective hydrogenation of acetylene to ethylene the activity of nickel can be diminished, and its selectivity increased, by 'selective' poisoning. Alternatively, the less active metal iron may be used. Palladium, however, exhibits both high specificity and high activity in this situation.

Other examples of metals particularly efficient in specific reactions are copper for the saturation of groups appended to the benzene ring, zinc for hydrogenation of aldehyde groups conjugated with olefine bonds, cobalt for double bond shift, and silver for the oxidation of ethylene to ethylene oxide. The copper-based ICI catalyst 52–1 is selective for the reaction carbon monoxide and steam to form carbon dioxide and water. ICI catalyst 51–1, also copper-based, is highly selective towards the hydrogenation of carbon monoxide to methanol.

(iii) *Activity*

In parallel with chemisorption, specific activity increases from right to left in each long period. It is relatively small in groups 1B and 2B but rises sharply between groups 1B and 8. For the hydrogenation of most unsaturated molecules (except nitrogen) there is a maximum of activity in group 8 at or near the Ni, Pd, Pt, or the Co, Rh, Ir triads. For ammonia synthesis activity is zero to the right of group 8, and increases monotonically on proceeding to the left in that group, becoming appreciable over cobalt and its congeners. Only in the decomposition of hydrogen peroxide are Cu, Ag, and Au more active than Ni, Pd, and Pt.

Obviously only the noble metals can be employed in acid media.

The activity of the metals is reduced or destroyed by adsorption, and by alloy or compound formation, when the processes give rise to a less active or inactive surface. The very reactive electronegative species (oxygen, sulphur, halogens) and the electron-pair donors (carbon monoxide, ammonia, water, hydrogen sulphide, phosphine, etc.) and their derivatives can be powerful poisons when present as adventitious impurities or as reactants, depending upon the facility of the catalysed reaction. These poisons may be adsorbed or they may react to form compounds in surface layers (oxides, sulphides, halides, etc.). If activity recovers on exclusion of the poison from the reactants, the poisoning is said to be reversible, otherwise it is permanent.

Mercury typifies the inactive metal which poisons by forming an inactive surface alloy.

Great care must be taken in catalyst design to ensure that proper weight is given to the inhibition of parasitic reactions. In dehydrogenation, for instance, it may not be desirable to choose the most active metal because of the occurrence of cracking and carbon deposition. A somewhat less active metal, or the use of selective inhibition, may ultimately provide the more satisfactory catalysis.

Copper, silver, and gold melt at around 1000°C, and iron, cobalt and nickel near 1500°C. Therefore surface and bulk Tammann temperatures fall within the ranges 150–350°C and 300–600°C for each group respectively. Thus, to attain and maintain suitable specific area the group 1B metals must always be supported, as must the transitional metals when used at temperatures above 250°C. The nobler metals of group 8 melt at rather higher temperatures, but they, also, are almost always supported for reasons of economy and ease of handling.

The efficacy of bases and acids (insulators)

The insulators are well represented by the solid oxides of the third period Na_2O, MgO, Al_2O_3, SiO_2, and P_2O_5, which exhibit the progression from basic, through amphoteric to acidic properties. A similar pattern holds for their congeners in the 4th, 5th, and 6th periods, although the higher members of groups 4 and 5 are neither electronic insulators nor very acidic.

The active solids are amorphous or crystalline, although the crystallites may be defective, and often contain a residue of combined water. The alkaline earth oxides are crystalline, but the lattice constant may be larger than observed in well-sintered material. Alumina derived from its hydrates is less crystalline and exists in a number of closely related structures formed by various stackings of the layers of aluminium-centred octahedra. Silica is frequently amorphous, and phosphorus pentoxide may be glassy.

Apart from the end-members (Na_2O, P_2O_5), which must in use combine with water, these oxides are all of high melting point and resistant to sintering.

The halides, especially the chlorides, of the closed-shell cations (e.g. $MgCl_2$) could be included among the insulators for convenience, but they are difficult to obtain and maintain in high area, and are not found in catalysts except as modifiers. The halides of higher valency, such as $AlCl_3$, are, of course, much used as Friedel–Craft's catalysts.

(i) Chemisorption

The exposed cations at the surface may remain bare, or combine with adsorbate and accept either a single electron or a lone pair from the adsorbate molecule or its fragments. None of the simple, non-polar molecules, or the smaller saturated hydrocarbons, are chemisorbed by the oxide insulators to an appreciable extent at low or moderate temperatures. Polar molecules are taken up, however. Water reacts to give hydroxyl groups and hydrogen-bonded forms, and alcohols yield among other species alkoxide-like complexes. Single electron transfers appear to occur only for complex molecules of low ionisation potential. Perylene, for example, forms a cation-radical on silica-alumina. A selective chemisorption of acidic species (e.g. phenol) occurs by bases, and of bases (e.g. NH_3) by acids.

(ii) Selectivity

Bases and acids are relatively ineffective for the redox reactions listed in fig. 2, although bases have significant activity for hydrogen transfer from alcohols to unsaturated organic molecules, and in the decomposition of alcohols tend to cause dehydrogenation rather than dehydration. This minor redox activity is sometimes of use in complicated reactions involving the more reactive types of molecules.

The typical activity is that of a general acid or base. The bare cation can function as a Lewis acid, with water as a Lewis acid and co-catalyst, or with dissociated water as a Bronsted acid, as, for example,

$$-\text{O Al} + RX \rightleftharpoons -\text{O Al} - XR \rightleftharpoons \left[-\text{O Al}\right]^{-} R^{+}$$

$$-\text{O Al} + H_2O + RCH=CH_2 \rightleftharpoons \left[-\text{O Al OH}\right]^{-} [CH_2CH_2R]^{+}$$

$$-\text{O Al OH} \rightleftharpoons \text{O Al O}^{-} + H^{+}$$

(where each Al is bonded to additional O atoms above and below)

In general, therefore, these basic and acidic solids catalyse reactions of the same kind as those affected by their homogeneous counterparts, but, unlike the aqueous acids or bases, a given solid possesses active centres of different types and strengths. The catalysed reactions most often encountered are those illustrated in fig. 2 under the headings of substitution, addition, elimination, and molecular rearrangements. The solid bases have not been much investigated, except in aldol condensations at temperatures at which they sometimes cause dehydration to follow the first step, but current research gives some suggestion of the occurrence of carbanion intermediates (ref. 3). The reactions brought about by the solid acids are best described within the framework of carbonium ion chemistry.

The specificity of the halides of the closed shell cations of large radius, in so far as this has been examined, lies largely in hydrohalogenation and dehydrohalogenation. The mechanism may involve polarisation of the adsorbates by the ions of the surface. The high acidity of cations of small radius

(BF_3, $AlCl_3$) is well known and can be employed to effect all the acid-catalysed processes under discussion.

(iii) Activity

Insufficient work has been done to allow an estimate to be made of the relative specific activities of the range of oxides from bases to acids. Among the acids it is well known that the acidity of silica is greatly increased by the addition of minor amounts of other oxides (in a state approaching solid solution) either to the bulk or the surface of the amorphous material. Hence the much-used silica-alumina, silica-magnesia-alumina catalysts, and latterly the molecular sieve alumina-silicate catalysts. Treatment with hydrofluoric acid or other compounds of fluorine can further increase the activity of the acidic oxides. Naturally the bases are poisoned by acids and vice-versa, and for this reason residual alkali, for instance, is usually kept to a minimum in catalysts for acid catalysed reactions, although small amounts may be added to control selectivity.

The alkali-metal and alkaline-earth metal chlorides are not as active as the chlorides of metals of variable valency in dehydrochlorination and oxychlorination.

The refractory binary oxides melt near or above 2000°C and can be prepared in high area modifications which are resistant to sintering. High area is more easily produced when the cation is of small radius and high charge, characteristics which correlate also with colloid and glass-forming tendencies and with very small solubilities in water. Consequently as a group they are substances most used to support or stabilise the less stable active phases; alumina is the substance which first springs to mind when a catalyst support is needed. Refractory compounds derived from two or more of the simple insulators are also effective but at the present time, with the exception of the cements, are seldom preformed.

The halides are of much lower melting point and higher solubility. They cannot be made or maintained in high area and are susceptible to attack by water so cannot be used as supports. Their chief function is to stabilise the more volatile, transitional metal halides by formation of solid solutions or compounds, e.g. KCl with $CuCl_2$ in oxychlorination catalysts.

The efficacy of semiconductors

Only the conducting oxides and sulphides, and some halides are of practical interest; the elements have little activity and, apart from graphite and 'carbon', no worthwhile properties as supports.

The active solids are more or less crystalline or amorphous and the bonds in the lattice vary from the ionic (halides) to the covalent (in some sulphides and volatile halides). In reducing systems only those oxides and sulphides which are irreducible can exist, that is to say the oxides of all metals except those discussed under metallic catalysts and most sulphides except those of the precious metals. Furthermore the non-stoichiometric excess of oxygen (or sulphur) required to confer p-type conductivity cannot be maintained under reducing conditions so that (see fig. 4) chromic oxide and maganous oxide become insulators or n-type semi-conductors. In oxidizing conditions the n-type semi-conductors tend to become stoichiometric but the p-types take up excess oxygen or sulphur.

(i) Chemisorption

The complexity of this process precludes anything but the briefest account of the descriptive aspects. The magnitude of the surface coverage by chemisorbed non-polar gases lies between the high values found for the transitional metals and the negligible quantities adsorbed by the insulators. Nitrogen is not known to be chemisorbed to any appreciable extent by oxides or sulphides, whereas hydrogen, the carbon oxides, hydrocarbons and oxygen are taken up to varying extents, depending upon the character of the solid. Little information is available concerning sulphides.

The division into n-type, p-type, and intrinsic semiconductors, or into high or low valency states is of some help. The simple gases and, presumably, other gases also are chemisorbed in at least two forms—reversible and irreversible. Thus chemisorbed hydrogen may be desorbed either as hydrogen or as water (or H_2S), and carbon monoxide as such, or as carbon dioxide, the more extreme reactions leading to some reduction of the surface.

Hydrogen chemisorption near 20°C by p-type oxides in their oxidised state is small, but is considerable in the case of n-type oxides (e.g. ZnO) in the reduced state. On zinc oxide, for instance, with increasing temperature, hydrogen is chemisorbed in several dissociated forms, probably involving oxygen-hydrogen bonds as well as metal-hydrogen bonds, as, for example:

$$Zn^{2+}(s) + O^{2-}(s) + H_2(g) \rightleftharpoons ZnH^+(s) + OH^-(s)$$

$$2Zn^+(s) \quad\quad + H_2(g) \rightleftharpoons 2ZnH^+(s).$$

Similar complexes no doubt form on the surfaces of other oxides. Hydrogen is probably adsorbed with some strength by all the oxides of the transitional elements except those with empty, half-full or full d-shells. In the case of adsorption on V_2O_3, Cr_2O_3, the surface is reduced by carbon monoxide

$$CO(g) + 2O^{2-}(s) \rightleftharpoons CO_3^{2-}(s) + V_{O^{2-}} + 2e$$

and can then adsorb oxygen to fill the vacancy (V) and trap the electrons (e)

$$\tfrac{1}{2}O_2(g) + V_{O^{2-}} + 2e \rightleftharpoons O^{2-}(s)$$

Oxides of p-type tend to chemisorb more oxygen than those of n-type, and the element seems to exist on the surface as O_2, O_2^-, O^- and O^{2-}, according to the temperature.

(ii) Selectivity

The oxides, sulphides, and halides catalyse in one way or another all the reactions listed in fig. 2, except the most characteristic acid-base reactions. It is evident that sulphides will be the active species in the presence of sulphur-containing reactants, and the halides in the presence of hydrogen halides, organic halides, and so on.

Under reducing conditions (e.g. in dehydrogenation) only the n-type solids and near-insulators (e.g. Cr_2O_3, MnO) can be used. In oxidation all types may appear, depending upon the chemistry of the system. It seems that the n-type oxides are the most selective in oxidation reactions. Compositions containing the oxides of sections C and F in fig. 4 recur in catalyst formulations.

A slightly reduced hydrous tungsten oxide is anomalous in that it provides an excellent dehydration catalyst.

(iii) Activity

Under reducing conditions the oxides of group 6 (Cr_2O_3, MoO_{3-x}, WO_{3-x}) are the most active in hydrogenation, dehydrogenation and dehydro-aromatisation. Over the sulphides, activity seems to peak near cobalt and molybdenum (ref. 4), so that 'cobalt molybdate' (ICI catalyst 41–3/4) is one of the most used hydrodesulphurisation catalysts. Little is known of the activity of ternary compounds except for compounds of zinc oxide with insulator oxides, where no hydrogenation activity is apparent unless the zinc ion is reducible (ref. 5). Transitional metal ions such as Cr^{3+} remain effective when dissolved in insulating solids (Al_2O_3) (Ref. 5).

High oxidation activity is confined to the p-type semi-conductors, and to oxides such as CuO. In hydrocarbon and carbon monoxide oxidation an incomplete activity series is roughly

$$Pt, Pd > Ag > MnO, Co_3O_4, CuO > NiO > Fe_2O_3 > Cr_2O_3 > V_2O_5 > ZnO$$

where the metals have been introduced as a comparison. Oxidation activity decreases with increasing atomic number in each group of the periodic table.

The halides most frequently found in oxychlorination catalysts are those of copper.

Composite catalysts

A catalyst is composite when it contains more than one chemical entity. The addition of a second component may be necessary to support or stabilise the active phase by a second and more refractory solid, or because the reaction is complex, involving a series of steps, each requiring selective catalysis. Almost all industrial catalysts are composites, if only because the use of a support decreases manufacturing costs and facilitates handling.

The rudiments of catalyst design rest upon the facts that the efficacy of a solid phase catalyst is determined by its selectivity, specific activity, and specific area, and by the effects of specific inhibitors. Thus, for a relatively simple process, such as the complete hydrogenation of an unsaturated molecule, the metal giving the best activity for the price is chosen, and it is also supported in a manner that will expose the highest possible area. If a high-area support can be found that also happens to possess hydrogenation activity (e.g. Cr_2O_3) so much the better, but the specific activity of the active metals is so great that attention is best directed towards supports that expose the highest area of the prime catalyst.

ICI catalysts 38–1 and 38–2, designed for the selective hydrogenation of acetylene to olefines, exemplify these principles. The active and selective metal is palladium, and the support is alumina, in which the pore structure has been so adjusted as to give combinations of activity and selectivity best suited to their specific requirements. In the same class is ICI catalyst 35–4, highly active for the hydrogenation of nitrogen to ammonia. It consists essentially of face-centred cubic iron crystallites stabilised with refractory oxides and it is promoted with alkali. This gives it resistance to sintering and poisons. Here the distribution of the non-metallic components is all important.

Nickel is the active component of the methanation catalyst (ICI catalyst 11–3) which removes CO and CO_2 from ammonia synthesis gas by hydrogenation to methane; again the supports and the promoters, and the method whereby these are compounded are vital to the final function of the catalyst.

The steam reforming of methane and naphtha over nickel-containing catalysts (such as ICI catalysts 57–1, 54–2, and 46–1) exemplifies the more complex process. In this reaction the metal dehydrogenates, demethanates and cracks the paraffins to low molecular weight olefines and catalyses the reactions of steam with methane, the olefines and their polymers. Finally it effects the carbon monoxide–water shift reaction. Carbon can be formed as a parasitic side reaction because nickel is only partially selective at these temperatures. Any residual acidity of the support can also cause carbon to form. The addition of alkaline species to the catalyst suppresses the formation of carbon by either mechanism. The lower hydrogenation—dehydrogenation activity of magnetite (Fe_3O_4) and of copper is associated with greater selectivity, and this is demonstrated in ICI catalysts 15–4/5 and 52–1, designed for the carbon monoxide–water shift reaction, where the alternative reactions of carbon formation and methanation are not catalysed.

As already mentioned alumina is frequently employed as a catalyst support. It has acidic properties and will catalyst reactions such as dehydration unless neutralised with alkali. Thus in the production of alcohols by the hydrogenation of aldehydes or ketones, the dehydration of the alcohols may occur unless alkali is incorporated in the catalyst. Other less basic oxides such as ZnO might be added, but the possibility of reduction to form less active alloys with the active metal must be taken into account. Depending upon the reaction temperature, this may be either an advantage or a disadvantage.

Similarly in exhaustive oxidation a very active precious metal may be extended on an inert support, or an active metal or metal oxide may be stabilised by the presence of another very active oxide (e.g. Ag/Ag_2O with oxides of manganese). Here the operating temperature greatly influences the selection of the appropriate combination.

The catalyst for a complex reaction must contain phases having activity appropriate to each step of the mechanism. The catalyst is then said to be multi-functional, and the intermediates diffuse between the phases, either through the gas phase or over the surface. A single phase can be multifunctional. Transitional metal oxides can dehydrogenate, decarbonylate, decarboxylate, and crack carbon-carbon bonds—but fortunately not usually with equal efficacy under any given set of conditions.

The isomerisation of saturated hydrocarbons affords a well studied example (ref. 6). Olefines are more reactive than paraffins and are rapidly isomerised by acidic catalysts. Paraffins can be rapidly dehydrogenated to olefines by platinum. Therefore at temperatures sufficiently high to give some dehydrogenation a catalyst composed of platinum metal dispersed upon silica-alumina should be more effective than the acidic catalyst alone. In practice this is true, and the reaction proceeds by olefine formation on the metal, diffusion of the olefine to the acidic phase where it isomerises, followed by diffusion back to the metal phase and rehydrogenation. The stationary concentration of the intermediate may be too small to be detected.

Some features of the hydrogenolysis of alcohols can be interpreted by a virtual mechanism in which the alcohol is first dehydrated to an olefine and the olefine subsequently hydrogenated over a metal. This reaction is therefore assisted by acidic supports and inhibited by alkalies.

In the oxidation and ammoxidation of olefines the hydrocarbon undergoes a partial dehydrogenation to give an adsorbed radical into which oxygen is inserted. The resulting aldehyde-like species may then condense with ammonia, and the adduct dehydrogenate, to a nitrile. The required functions—dehydrogenation, oxygen insertion and condensation—are associated with catalysts such as bismuth molybdate and tin oxide-antimony oxide compositions.

The properties of the component phases must clearly be chosen so that they match both the activity and the selectivity of the mechanism. If a reactive intermediate is produced at a faster rate than that at which it can be consumed, then undesirable parasitic reactions will ensue. The components must also be compatible. They must be so selected as to avoid *in situ* reactions which would alter their chemical and catalytic characteristics.

Conclusion

This brief survey of the properties of solid catalysts has demonstrated that they can sometimes, and in theory, be tailor-made for particular tasks. Much more could be written on all of the themes here dealt with. But despite the fairly elementary level of this chapter, and despite the fact that an empirical approach must still, for the time being, largely govern the preparation of catalytic solids, an application of the broad principles just outlined should increase efficiency.

CHAPTER 2

The structural engineering of catalysts

A SUCCESSFUL commercial heterogeneous catalyst not only must be capable of catalysing the desired reactions selectively—it must be also mechanically robust. It must be suitably shaped, so that the reagent fluid or gas can flow through a bed made of it without excessive pressure drop or uneven distribution. And finally it must retain both its reactivity and its mechanical properties over a

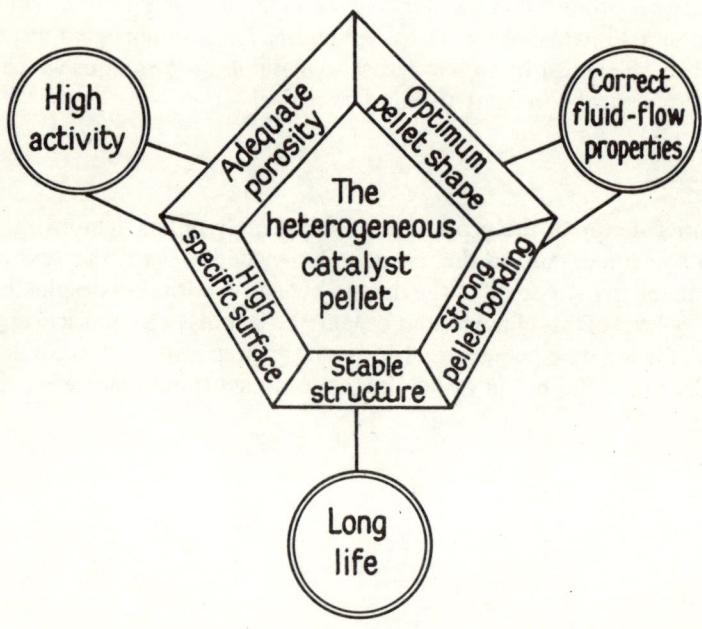

FIG. 5. Requirements of an ideal catalyst.

long life, which includes start-ups and shut-downs. These three requirements of high activity, correct fluid flow properties, and long life, together dictate a number of structural requirements (fig. 5). In order to have high activity (either per unit weight or per unit volume of catalyst) it is desirable that the catalyst has a high specific surface, and that the catalyst is adequately porous so that the reactants can gain access to the inner surfaces of the catalyst pellets by diffusion. Ideally, for a catalyst which is strongly internal diffusion limited, a porosity of 50 per cent gives the highest activity per unit volume with the catalyst in its operating condition (i.e. reduced, if it is a metal). It is usually difficult to attain a porosity as high as this in a particle of adequate strength, and part of the science of catalyst formulation is concerned with an attempt to reconcile these conflicting demands. The requirements of high strength and a stable structure depend on a firm welding together of the catalyst components into a structure which is not greatly weakened or changed by sintering during use. A final requirement for a successful commercial product is that it should be reproducible in its behaviour, and relatively cheap.

High specific surfaces

The need for a high specific surface follows immediately from the fact that reactions are catalysed at the surface. When, as is commonly the case, the catalyst is a solid, a high specific surface is obtained by preparing the solid in the form of very small crystallites having sizes in the range 50 to 5000 Å. The results of a typical experiment performed as part of a study of the relation between catalyst activity (in the absence of diffusion limitations) and the total exposed surface area of the active species in a multi-component shift catalyst is shown in fig. 6. The selective chemisorption of oxygen on the surface of copper was used to measure the exposed copper surface, and the copper crystal-

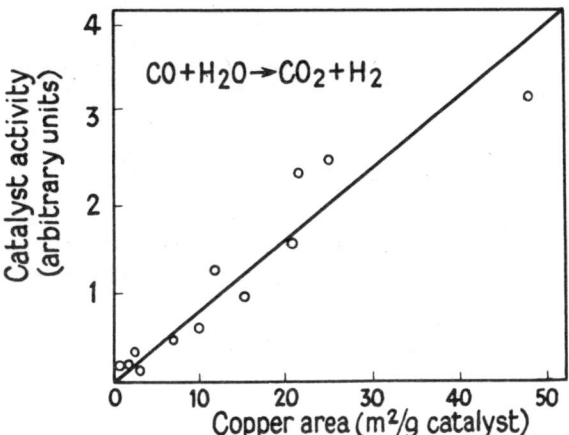

FIG. 6. Relationship of catalyst geometry to catalytic activity. Copper crystal sizes and exposed surface areas measured by X-ray diffraction and oxygen chemisorption. (E. Thurston, to be published.)

lite sizes were measured by line broadening in the X-ray diffraction image. Good agreement was found between these two techniques for assessing the copper geometry. In commercial practice most catalysts operate in the internal diffusion control regime where the relation between activity and specific surface is not linear (as with pure kinetic control) but is broadly proportional to the square root of the specific surface, and also to the square root of the particle voidage.

The influence of internal and external mass transfer in catalyst design

Fig. 7 has been drawn in order to show, though only roughly, the quantitative relations between the main variables relating the catalyst particle's activity to its structure. This figure has been calculated for gaseous reactants and

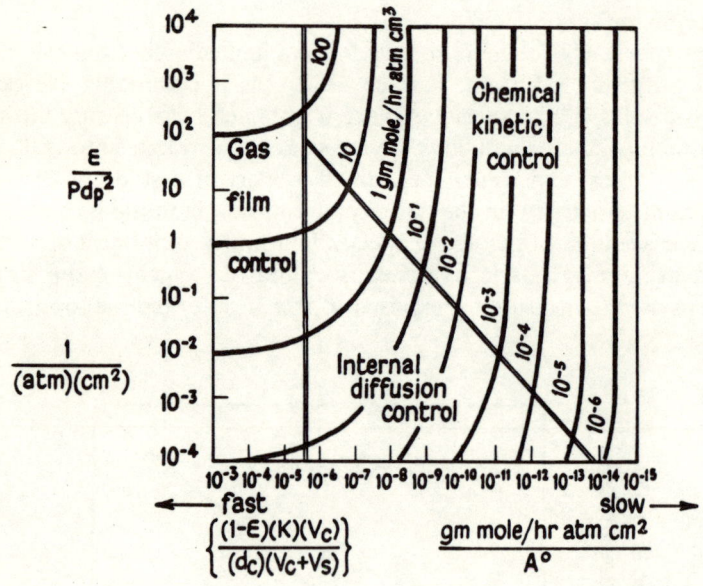

FIG. 7. Relationship factors determining catalyst structure. ε, voidage; V_c, volume of catalytic species; V_s, volume of support; K, specific catalysis rate constant; d_c, catalyst crystal diameter (Å); P, pressure (atm); d_p, catalyst particle diameter (cm).

products, assuming typical values for the gas film mass transfer coefficient and for the various diffusion coefficients. The inherent ability to catalyse is the product of the specific rate constant (K), the specific surface of active species (proportional to $1/d_c$), and the loading of the active species $(1-\varepsilon)\{V_c/(V_s+V_c)\}$. At high values of these terms (that is, with a high ability to catalyse) the reaction is limited by the rate of transportation of reactants and products to and from the catalyst (i.e. by gas film control). Lower values give either internal diffusion or chemical kinetic control. The former operates when catalyst particle voidage is low, particle diameter is high, or pressure is high (and hence the diffusion coefficient is low). The latter occurs under the reverse conditions. Many catalysts fall in the internal diffusion control region, some bordering on gas film control whilst others border on chemical kinetic control. The highest overall kinetic constants (marked on the curves) occur in the top left hand of the figure, showing the desirability of a high ratio of catalytic material to support (V_c/V_s), a low crystal size for the catalyst (d_c), a low size for the catalyst particle (d_p), and an optimum voidage (ε) in the catalyst particle.

The structural properties of catalyst components

Considered as a structure, the major factor that can cause loss of surface area in a catalyst is crystal growth through diffusional merging of contiguous smaller crystals. Simple solids like metals or metal oxides sinter particularly

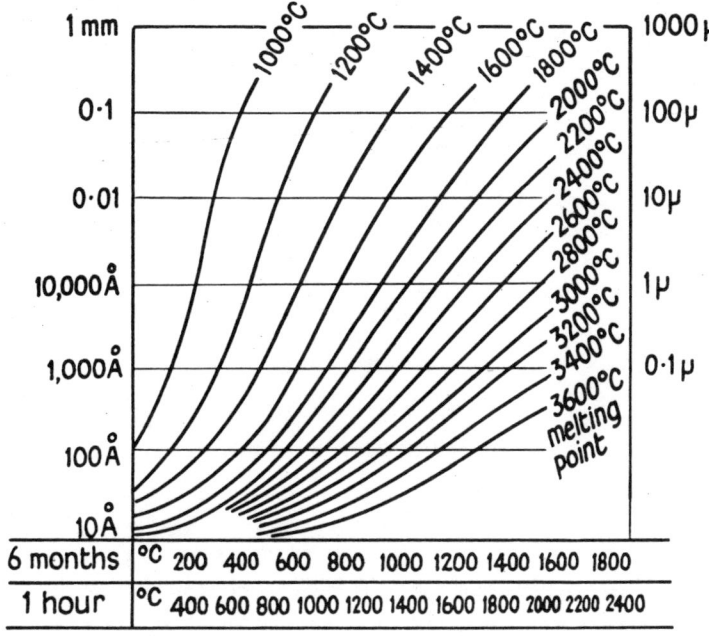

FIG. 8. Minimum crystal sizes for sintered compacts.

readily if they are in the form of very small crystals (< 500 Å)—especially at temperatures above about one half of their melting points. Fig. 8 gives a general indication of the relation between melting point, sintering time, and the minimum crystal size that could exist in a single component compact after sintering. From fig. 8 it can be seen that, for instance, if a compact is made of copper (m.p. 1083°C), and this is sintered at 200°C for 6 months (in a reducing atmosphere), then the minimum crystal size will be over 1000 Å, and if sintered at 300°C will be over 1 μm, whereas alumina (m.p. 2032°C) could be held at 500°C for 6 months without its crystals growing to over 70 Å. For this reason catalysts in which the active species is a metal having a relatively low melting point, normally also contain much more refractory crystals which act as 'spacers', and stop the readily sinterable metal crystals coming in contact with each other. Alumina, chromia and magnesia are very commonly used as such 'stabilisers' of the finely divided state of an otherwise readily sinterable catalytic species.

In addition to the catalytic species and the stabilising species, a catalyst formulation often also contains a further inert species, the support, which gives body and strength to the catalyst granule. The support also decreases the concentration of the usually more costly catalytic species. A ceramic oxide compound is frequently employed, with or without the addition of cementing materials.

The overall catalyst formulation is thus likely to consist of three main parts (fig. 9), the catalytic species, the stabiliser or dispersant, and the support or

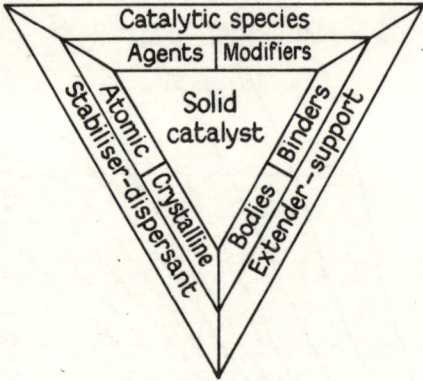

FIG. 9. The elements of a solid catalyst.

extender. Each one of these parts may be complex in nature. Thus the support may be composed of a ceramic body cemented together with a hydraulic cement. The catalytic species may consist of a metallic element as the main catalytic agent, the surface of which is modified by the addition of a promoter or modifier. Stabilisers also can be complex and multicomponent, existing either in the form of gross crystals, easily discernible by electron microscopy, or else in a highly dispersed or nearly atomic state.

Geometric and strength factors in catalyst structure

Geometric factors

Although the details of the optimal structural design of a catalyst vary with the particular system and chemistry involved, one general rule applies fairly widely. This is the rule expressing the effect of the geometrical relationship between the average crystalline size of a well-dispersed refractory stabiliser, the volume ratio of sinterable material to stabilising material, and the average crystal size of the sinterable material after sintering. When distribution is good, without segregation of the two constituents, and the sinterable material has small initial size, experiment indicates, very roughly indeed, a relationship of the type shown in fig. 10. The smaller the size of the dispersing material, the smaller the size of the sinterable material after sintering; and the lower the fraction of sinterable material the lower its crystal size after sintering. Catalysts tend to be very varied in the ratio of sinterable component volume to dispersant

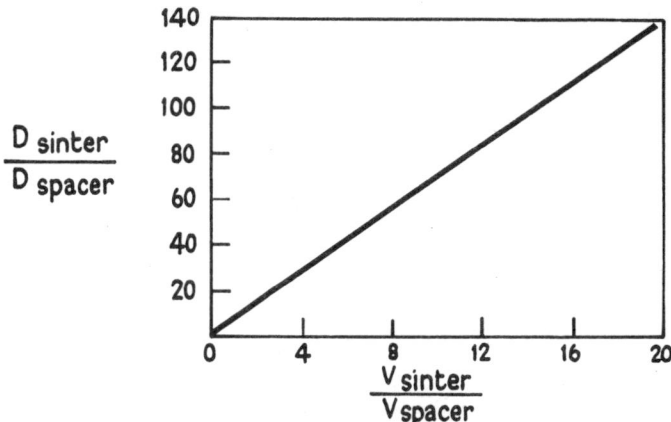

FIG. 10. Stabilisation of sinter size by high-area supports.

volume—from ratios of less than one up to ratios of well over ten. The crystal size of the dispersion material is often very small—around 20Å in some formulations.

Strength factors

One of the commonest methods of forming catalyst granules is by pelleting in a tabletting press. This method has the advantage of producing very regular, highly polished cylinders or rings, which flow readily during packing of the catalyst bed, and hence pack to a uniform voidage, giving uniform fluid flow properties and good distribution. Pelleting in a press can also give a hard, strong granule, capable of standing up to both the superimposed load during use, and the often greater impulsive loads occurring during catalyst charging. Pelleting is, however, capable of producing these desirable properties only if the tabletting machine punches are correctly shaped, and the material of the feed to the machine is correctly sized and of the optimum density. Furthermore, enough (but not too much) die wall lubricant must be added to the feed to reduce wall friction. This reduction in wall friction spreads the pelleting pressure, exerted by the top and bottom punches, more uniformly through the pellet, resulting in uniform compaction and absence of differential stress, which otherwise might relax by fracturing the ejected pellet when the pelleting load is removed. Typical examples of such stress effects are capping and lamination of the pellet. Die-wall lubricants commonly used are graphite and calcium stearate. An excessive quantity of lubricant in the catalyst weakens the structure.

No doubt all solids are pelletable in theory, if subjected to sufficient triaxial stress. But in practice, catalyst pelleting is performed in a machine constructed of steel punches and dies. The stresses applied must therefore be well below those which would cause permanent deformation to the punches. And so the main pelletable component must be substantially weaker than steel, and must flow plastically before the steel flows. Now, there is a broad general relationship between the strength and hardness of inorganic solids and their melting points (fig. 11). Low-melting-point solids have low hardnesses and strengths, and high-melting-point solids have high hardness and strength. The general sequence

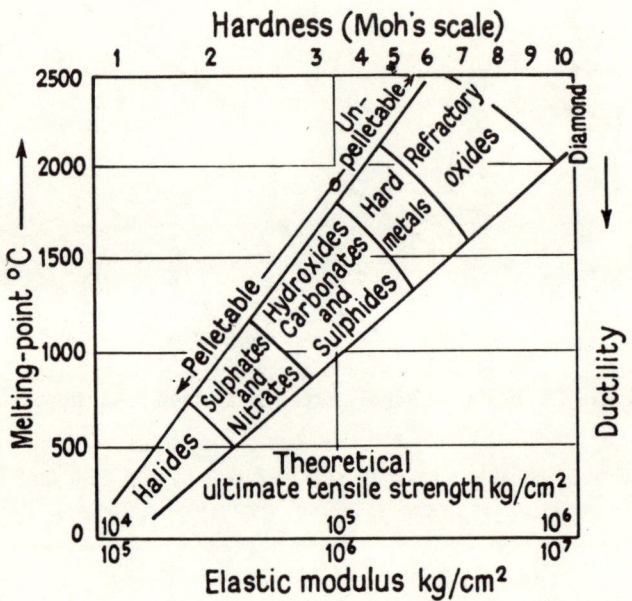

FIG. 11. Relation between strength, hardness, and melting-points for inorganic solids.

is halides, sulphates–nitrates, hydroxides–carbonates–sulphides, hard metals, refractory oxides. Only the weakest three categories, up to a Moh's hardness of about 4, can be satisfactorily pelleted in a normal tabletting machine. This means, for instance, that a feed powder consisting only of a mixture of Al_2O_3 and graphite lubricant could not be pelleted. To make the mixture pelletable a third component has to be added which conforms to the pelletability criterion.

A catalyst pellet must be sufficiently strong to resist, without appreciable breakdown, four types of stressing. Firstly, it must be strong enough to resist abrasion during transit due to the rolling and tumbling to which it is subjected inside the container drum. Secondly, it must be strong enough to resist the impact loading experienced during the charging of a converter (when it may fall several feet on to a pile of material already present). Thirdly, it must have sufficient internal cohesion, and be so formulated, that gross chemical changes (such as reduction or oxidation of some of the pellet constituents) can occur whilst the catalyst is in use, without these changes causing pellet break-up. Fourthly, the reduced pellet must be capable of withstanding the load in the catalyst bed due to gas pressure-drop, weight of superimposed pellets and, on occasions, the relative movements caused by thermal expansion and contraction of the bed and its containing vessel. The capacity to withstand the first and second of these loads is often directly related to the tabletting operation, since the composition and structure of the catalyst often remains unchanged from the completion of tabletting until the occurrence of reduction in the plant reactor. Some catalysts, however, undergo further treatments following tablet-

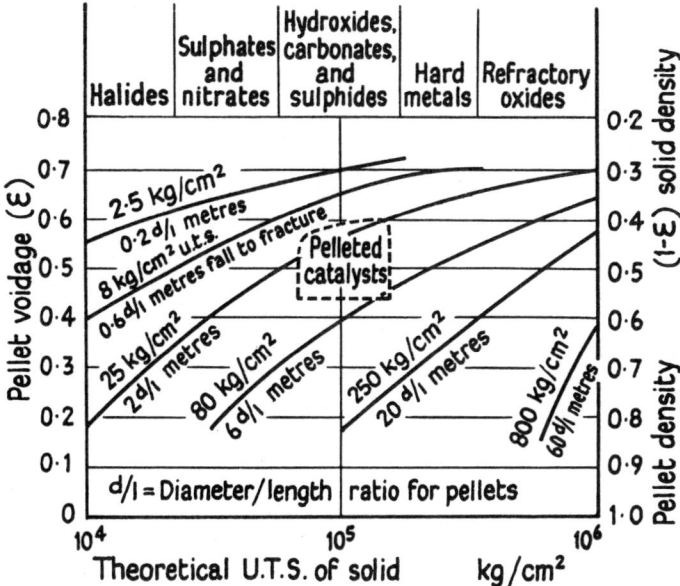

FIG. 12. Relation between strength requirements and pellet voidage and pelletability.

ting, such as impregnation, calcination, or cement setting. In these catalysts the final pellet strength is less directly related to the tabletting operation.

The impact loading during charging is generally much greater than the loading suffered during transit, and the strength of pellets under this type of loading can be described in terms of the minimum drop height producing failure. It is, of course, necessary to specify the nature of the surface on to which the catalyst drops. This surface is ideally of the same hardness as the catalyst. It can be harder without greatly altering the result, but it must not be softer. Within these limitations, the minimum drop for failure may be related to the pellet length/diameter ratio, and to pellet ultimate tensile strength (UTS) by the roughly approximate relation

drop for failure (metres) ≈ 0.075 (d/1) (UTS)
UTS is expressed in kg/cm^2 or
drop for failure (feet) ≈ 0.02 (d/1) (UTS)
UTS is in lb/in^2

From this expression it will be seen that long thin pellets are much weaker under impact loading than short, fat ones. Very commonly d/1 $\approx$ 1, and pellets are designed to have a minimum drop of about 3 m or a UTS of 40 kg/cm^2 (570 lb/in^2).

Fig. 12 relates the strength requirements of a pellet with the pellet voidage, and the pelletability properties dealt with in fig. 11. For the many catalysts which are diffusion limited, voidages of between 0·4 and 0·6 are desirable which, coupled with the UTS requirement of about 40 kg/cm^2, and the upper limit of hardness set by the tabletting machine construction, limit the zone in which satisfactory raw pellets can be made by tabletting. This is the area indicated in the middle of fig. 12.

Catalyst poisoning and structural collapse

As already stated, catalysts can be poisoned in use in many ways. Here we are particularly concerned with poisoning resulting in a collapse of the high surface area structure of the catalyst, with consequent loss of activity per unit weight, though not necessarily loss in specific activity (activity per unit area of active phase). A simple example is a porous two-component structure consisting of a sinterable and non-sinterable crystal aggregate. Two methods by which crystal growth could occur seem plausible. The first is the loss in stability of the non-sinterable component, which, under the influence of a changed chemical environment, starts to sinter. Fig. 10 shows how the crystal size of the highly sinterable material, now that it is no longer stabilised, grows at a rate proportionate to that of the less sinterable material. The action of water and steam on refractory oxides like alumina is an example of this weakening of the stabiliser. The second way in which crystals of the sinterable component crystals may grow is through the existence of a bypass transport mechanism. This can carry atoms of the sinterable component across the gap between one crystal of this component and another, thereby permitting the thermodynamic potentials of different size crystals to become effective driving forces in promoting crystal growth. Under these circumstances the stabilising support crystals need not grow, but the relationship illustrated in fig. 10 breaks down, and the laws governing the sintering of the sinterable material virtually (though not exactly) revert to those of the single component system illustrated in fig. 8. A good example of such a structural collapse mechanism is the effect of the presence of a little chlorine (or some other halogen) in a copper catalyst—even one well stabilised by alumina. The halogen forms a low melting point and somewhat volatile surface compound with the copper, readily enabling it to 'jump the gap' of tens of Angstrom units separating copper crystals, and the copper crystals grow from, say, 100 Å to 10 000 Å, in a period of a relatively few hours at temperatures as low as 200°C. Poisons of this type are much more destructive of catalyst structure than is high temperature and, because the structure is destroyed, the catalyst cannot be wholly regenerated even if the poison is removed.

Other constructional processes in catalyst building

The short space of this chapter is quite inadequate for a description of the many constructional processes used in catalyst building. But in a book discussing (amongst other catalysts) that for ammonia synthesis, mention must be made of the method by which the whole of the catalyst structure is formed during the very last stage of the catalyst fabrication process—that is, during the final reduction in the converter. Ammonia synthesis catalyst in its oxidised state is, essentially, a microstructureless aggregate of large crystals of magnetite, containing a solid solution of alumina. Other promoters are also added, but

FIG. 13. A reduced ammonia synthesis catalyst. Stereoscopic electron micrograph.

it is the alumina which is the chief non-sintering component. On reduction of the magnetite, which proceeds from the outside of the catalyst particle and moves inwards, the alumina is left behind along a network of crystal planes of the large magnetite crystals, each of which reduces to form a large number of small iron crystals held loosely in the alumina crystal network—the total solid volume of iron being markedly less than that of the magnetite from which it was formed. An electron microscope photograph shows the iron crystallites lined up like stacks of eggs in a crate, with very finely divided alumina and empty channels between the iron crystals (fig. 13).

In addition to pelleting, catalysts may be formed by granulation and by extrusion. In both these forms, various types of cementing agent are frequently used to give adequate strength to the finished catalyst. Each of these processes has its own science, so that a composition which is suitable for say, pelleting, is unsuitable in general for granulation or for extrusion and vice versa. As was emphasised earlier in this chapter, the construction of a successful commercial catalyst involves a compromise between many conflicting requirements. The general principles are clear, but their detailed application requires the bringing together of knowledge from catalytic theory, inorganic structural chemistry, colloid chemistry, materials technology, rheology, and chemical engineering. All are relevant in this very complex science.

CHAPTER 3

Catalyst testing

THE major objective of the industrial catalyst manufacturer is to produce a material which will catalyse a particular reaction on the commercial scale. The scale of factory operation, where sometimes hundreds of tons of catalyst are installed, and where the lifetime of some of the catalysts can be measured in years, makes it expensive and time consuming to develop catalysts by full scale testing. Small-scale catalyst testing procedures have therefore been devised in order to minimise the expense and time involved in obtaining data for innovation and development. It is important to bear in mind that there can be important effects of scale, both in size and time, on the performance of catalysts, so it remains at least partly true to say that the ultimate test is to find out how a catalyst works in the full-scale plant.

Nevertheless, provided a good understanding can be obtained about the reaction mechanism, it is possible to examine most catalysts satisfactorily under laboratory or semi-technical conditions. It is generally true that the cost and sophistication of the experimental techniques is fairly closely paralleled by their usefulness in predicting plant performance.

The development of new catalysts involves the testing of relatively large numbers of different formulations. The control of plant production also results in large numbers of samples for testing. Both of these operations require a rapid and reliable test procedure. For design work it is essential to be able to predict how potential catalysts will behave under plant conditions, and test techniques are available which simulate, to a greater or lesser degree, real conditions. Development then proceeds through semi-technical testing to full-scale trial. ICI is relatively fortunate in having a large number of operating plants where this final stage of development is possible. The steps involved in developing a successful catalyst are shown in fig. 14. The procedure is designed to permit the maximum number of variables to be examined in a way which results in all potentially suitable catalysts receiving much more detailed investigation. By this technique, full-scale testing is reserved for those catalysts which have passed the other tests.

A rapid initial screening is carried out at atmospheric pressure, so that those formulations showing promise can then be tested more rigorously under semi-technical and sidestream conditions, if necessary at pressure, to obtain basic kinetic and life data. Simultaneously, more detailed kinetic and poisoning information is obtained under differential operating conditions. Life-testing in plant sidestream units is regarded as an essential step, since these are the only conditions under which the catalyst is exposed to 'real' gases, which may, for example, contain trace impurities. Formulations which pass sidestream tests satisfactorily, over an extended period of time, are subsequently submitted to full-scale plant tests.

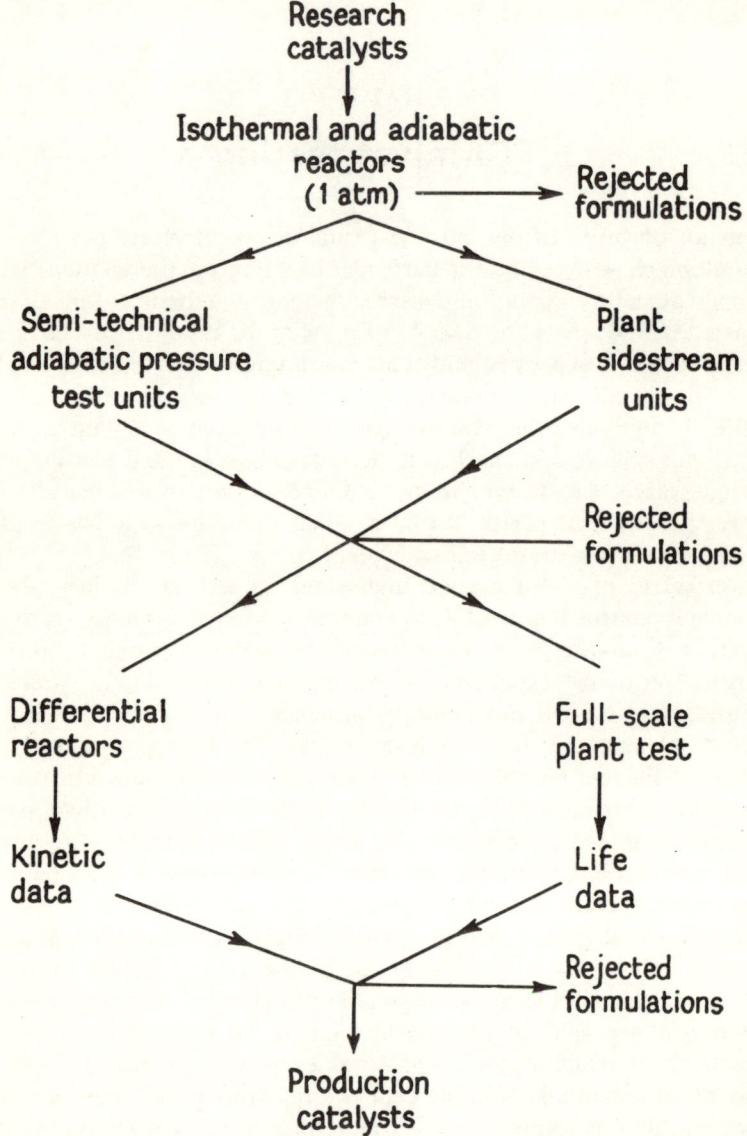

FIG. 14. Steps involved in developing a successful catalyst.

In addition to activity tests it is essential that new catalysts are tested for physical robustness. This is necessary in order to ensure that the catalyst particles are sufficiently strong to withstand shipment and charging. The catalyst particles must also be strong enough to resist any stresses that may be applied during use; caused by reduction, the gas flow, pressure drop and thermal cycling.

Another tool for controlling catalyst production, which can sometimes also be used for an assessment of catalyst activity, is the micro-meritic analytical technique. This enables the area and pore-structure inside the pellet to be measured. Refinements of the basic technique allow measurements to be made on the surface properties of the active material, sometimes under reaction conditions. These techniques, which are widely used in catalyst research, have been well described by many authors (ref. 7).

Catalyst activity

The performance of a catalyst is generally assessed in terms of the rate at which it promotes a desired (or sometimes an undesired) reaction. Reaction rate can be expressed by the well-known proportionality (see Chapter 1, p. 2):

$$\text{reaction rate} \propto e^{-E/RT} f(P, V, X).$$

The proportionality is termed 'catalyst activity', thus

$$\text{reaction rate} = \text{activity} \times e^{-E/RT} f(P, V, X).$$

At equivalent temperatures, pressures, reaction volumes and mole fractions of reactants, reaction rate is proportional to catalyst activity. Where other reactions are also possible, an assessment must also be made of catalyst selectivity—the ratio between activity for desired and undesired reactions. With most catalysts used for the production of synthesis gas and ammonia this is unimportant because, by careful catalyst selection, high selectivity is achieved under normal operating conditions.

Before considering how catalyst activity is measured, the properties controlling mass transfer to and from the catalyst surface are briefly examined. The flow of reactants to the surface of the catalyst can, and usually does, affect the measured activity of the catalyst. Specific activity is the measure of the reaction rate available when there are no external limitations. The term activity is used to describe conditions even when there are limitations; perhaps 'apparent activity' would be a more realistic term but 'activity' has become a generally accepted word covering both 'specific' and 'apparent'.

Rate-limiting steps in gas–solid catalytic reactions

Before a meaningful activity test on a catalyst can be made it is necessary to understand the mechanism by which the reactants are transported from the gas phase to the active surface of the catalyst, and by which the products are returned to the bulk of the fluid. Unless this is fully understood, and the rate limiting step or steps are recognised, serious errors may be introduced when the results are applied to full-scale converters. The commonly accepted route by which the reactants reach the active surface of a catalyst in pellet form is shown diagrammatically in fig. 15. Before reaction can take place the reactants in the bulk of the gas must be transported to the active surface of the catalyst. The first stage in this process is normally regarded as the diffusion of reactants to the surface of the pellet through the stagnant gas film which surrounds it. Having reached the pellet, the reactants have still to pass through its internal structure, consisting of a macro, micro and sub-micro pore structure, before they reach the active surface. The reaction products, desorbed from the catalyst surface, return to the bulk of the gas stream by a reverse procedure. Any limitation associated with transport to or from the active surface can lead to a lower

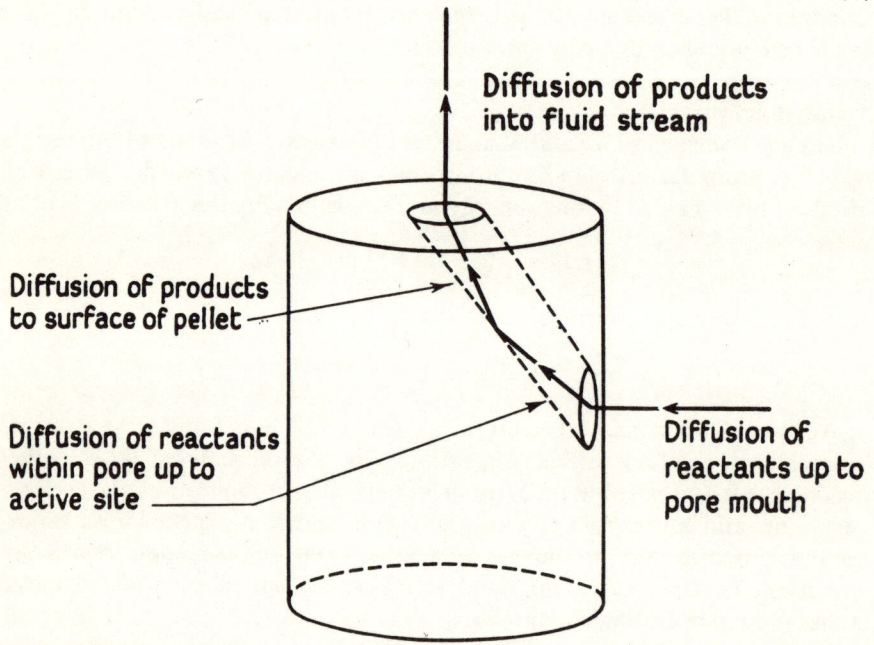

FIG. 15. Diffusion of reactants and products within catalyst pellets.

concentration of reactants or a higher concentration of products inside the pellet than in the bulk gas stream. For this reason activity measurements reflect an overall effect, and can sometimes be very different from measurements of either specific activity or activity under plant conditions.

With most catalysts used for hydrogen and ammonia production the overall reaction rate can be considered to depend on three separate processes:
(a) the diffusion of one of the reactants from the gas phase to the surface of the catalyst particle,
(b) the diffusion of a reactant in the pores of the catalyst particles,
(c) the rate of reaction at the active surface of the catalyst.

The extent to which either pore or gas film diffusion limits the rate of reaction increases when the specific rate of reaction on the catalyst increases. Diffusion limitations therefore increase with increased specific activity. Increase in temperature increases the likelihood of diffusion limitation, since the rate of reaction increases exponentially with temperature, whereas the rate of diffusion only increases as the $\frac{3}{2}$ power of absolute temperature. When the order of reaction is greater than zero, the rate of reaction increases with pressure, consequently increasing the degree of diffusion limitation.

Gas-film diffusion limitation

Normally the difficulty of mass transfer inside the pellets outweighs that of transfer from the gas stream to the pellet surface. It is only at high overall reaction rates (that is, with very active catalyst pellets or at high temperatures), and with low gas velocities, that gas-film diffusion may become limiting. It is worth remembering that this type of limitation is more likely to occur at gas velocities used for catalyst testing than under plant conditions, where reactor height usually ensures that velocities are also high. Care must be taken, therefore, to find out if testing is being carried out in this region, and, if so, steps must be taken to avoid it. Under a gas film diffusion limitation, the rate of reaction will increase with increasing gas velocity—a fact by which the presence of the phenomenon can be recognised, so that, if necessary, a suitable correction can be applied when calculating the gas velocity required on scale-up. Several experimental methods are available for checking on the possibility of gas diffusion limitation, but the one that has been found most appropriate for tubular reactors is to measure conversion at constant space velocity, but different gas rates—that is, employing different catalyst volumes. An increase in conversion with gas rate signifies gas film control. In cases where the measured activation energy is abnormally low ($\leqslant 3$ kcal/mole) a gas film diffusion limitation should be suspected.

The possibility of gas film control can also be roughly estimated by calculating the magnitude of the concentration gradient of the reactants between the bulk fluid and the surface of the catalyst particle. If the concentration difference is more than a few per cent of that in the bulk fluid, the presence of gas film limitation is likely. The concentration gradient will depend on the nature of the reactants and the degree of turbulence. The effect of turbulence varies for different systems, but experimental information is available for assessing its importance. The procedure has been simplified by Wheeler (ref. 8), who has shown that for a completely diffusion limited reaction

$$k_\infty = 10 \sqrt{\left(\frac{V_L}{\overline{M} a^3 P_T}\right)}, \qquad (1)$$

where k_∞ = first-order rate constant/unit volume reactor (s^{-1})
V_L = linear velocity of gas through empty reactor (cm/s)
$\overline{M}$ = average molecular weight of gas
a = pellet size (cm)
P_T = total pressure (atm)

Thus in order to be sure that a particular system is not film diffusion limited, the measured first-order rate constant should be less than 10 per cent k_∞.

Pore diffusion limitation

The effect of pore diffusion was discussed briefly in Chapter 2, where it was shown to be an important general limitation in the performance of most porous catalysts at moderate temperatures and, unlike gas-film diffusion, it is independent of the gas velocity outside the catalyst particles. On the other hand, for obvious reasons, it will be dependent on pore volume, pore radius, and particle size. A good treatment of the effect of pore diffusion on reaction rates has been presented by Thiele (ref. 9) and also by Wheeler (ref. 8). This treatment can be used to show that, for a strongly pore-diffusion-limited first-order reaction, the rate of reaction per unit catalyst volume is given by the equation

$$\text{Rate} \propto (k'D)^{\frac{1}{2}} \frac{(S_v V_v)^{\frac{1}{2}}}{d} C_i, \qquad (2)$$

where k' = first-order specific rate constant/unit area catalyst.
D = diffusivity
S_v = catalyst surface area/bulk volume
V_v = voidage/bulk volume
d = pellet diameter
C_i = reactant concentration

This equation shows, for a pore-diffusion-limited reaction, that the rate of reaction is dependent on the properties of the catalyst particles, for example, size, voidage, and surface area. It is therefore important to make allowances for pore diffusion limitations when the size of the catalyst particles and their cost are optimised. Potentially serious design errors might occur if the properties of the catalyst particles (size, for example) were changed during the scale-up to plant operation without regard to the effect on the reaction rate. The best experimental method for assessing the importance of pore diffusion is to measure the rate of reaction with varying particle size—in the absence of pore diffusion the reaction rate will remain unchanged.

As has been shown earlier, increase in temperature and pressure both increase the possibility of a diffusion limitation. Any indication of reaction rate being limited by pore diffusion in a system tested at moderate reaction temperatures, and at atmospheric pressure, will therefore become more pronounced at higher temperatures and pressures.

Gas distribution

In addition to the problems associated with the transfer of material to and from the catalyst surface, a further important practical problem in test reactors (and potentially in full sized converters) relates to differences in packing density that occur in the catalyst bed. This is because uneven packing will result in uneven gas flow rates in different parts of the catalyst bed, more gas flowing through those parts of the bed with high voidage than through the parts of lower voidage. The flow pattern of the gas through the test converter must be known before the results from a test bed can be applied to a full-scale plant, otherwise serious errors may result. The gas distribution problems discussed in this section relate to tubular converters, and so apply to the majority of test equipment and full scale units. The special conditions applying in stirred tank reactors are discussed in a later section.

Where gas flows through an empty pipe at a low Reynolds number (that is, at a low linear velocity) the velocity profile perpendicular to the flow is not uniform, but varies from a maximum at the centre of the pipe to zero at the wall—the profile is parabolic. At high Reynolds numbers, with increased turbulence, the velocity profile becomes more nearly planar and 'plug flow' results. Plug flow means that, over any cross-section normal to the direction of flow, the mass flow-rate and gas properties are uniform. This is an ideal condition for catalyst testing, and an additional advantage is that diffusion relative to flow rate is very small. The design equation for a tubular reactor, based on the assumption of plug flow, is

$$\frac{1}{\text{(space velocity)}} = \frac{V}{F} = \int_0^{x_1} \frac{dx}{r},$$

where V = catalyst volume
F = flow rate (moles/time)
x_1 = conversion
r = rate of reaction, moles per unit volume catalyst per unit time.

Because the design equation for a tubular reactor assumes plug flow, any deviations from this ideal state will lead to errors, and since plug flow is never attained in a packed bed, the importance of these errors must be assessed. Deviations from plug flow through packed catalyst beds result in gas distribution problems, and can arise either because of increased voidage near the walls of the converter or because of uneven packing in the bed. The former is more likely to be important with test reactors which often have a relatively large wall/volume ratio. The latter can occur in full scale plant whenever incorrect charging results in non-uniform catalyst packing.

The magnitude of the increase in voidage in the vicinity of the walls of a tube has been estimated by Roblee et al. (ref. 10) and by Schwartz and Smith (ref. 11). The effect this voidage difference has on conversion can be very considerable, especially under conditions where conversions are high and are not limited by chemical equilibrium—for example, in methanators and desulphurisers.

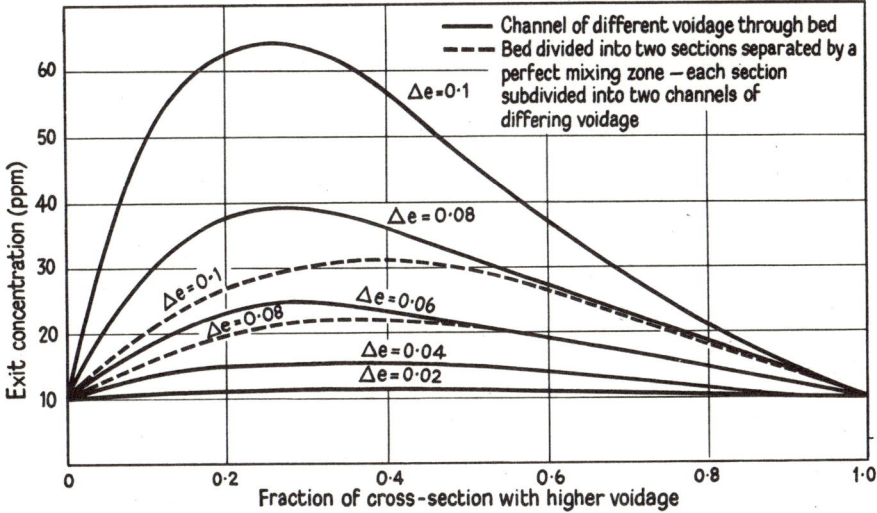

FIG. 16. Effect of voidage differences on apparent degree of conversion.

Craven (Chapter 6) has estimated the importance of this effect, using as an example the methanation reaction. He first considered the bed as being divided into two sectors, so that flow was divided into two parallel paths with different voidages. The difference in voidage between the paths, Δe, was varied from zero to 0·1 and the relative areas of the sectors was varied from being all of low voidage to all of high voidage. (When $\Delta e = 1·0$, this indicates that one path had voidage $e = 0$, and the other path was empty with $e = 1·0$. Δe can therefore range from 0 to 1·0.) The results are summarised in fig. 16 which shows how the conversion of 1 per cent CO with hydrogen over a nickel catalyst can vary depending on voidage differences in the bed. The system was set up so that, with an adiabatic converter, when there were no differences in the voidage between the sectors, the concentration of carbon monoxide was reduced to 10 ppm. The values of Δe can represent the wall effects obtained in tubes of different diameters, where the region adjacent to the wall has a low voidage, and a 4·8 mm ($\frac{3}{16}$ in) methanation catalyst in a 5 cm (2 in) diameter tube can result in the exit concentration of CO from the tube being six times higher than it would have been from an ideal converter (that is, plant scale converter).

The effect of uneven packing in a full scale converter is similar, except that the difference in voidage is unlikely to be in the same sector for the entire length of the converter. This effect has been simulated by extending the previous calculation to cover a bed made up of two longitudinal sections, each separated by a zone of perfect mixing. Each section is divided into two sectors with different voidages. These results are shown by the broken curves in fig. 16. Under extreme conditions, effects of this magnitude have been obtained in plant converters, which emphasises the need for careful catalyst charging.

Test reactor types

There are three factors which must be considered before deciding which type of test reactor should be used. These are the purpose for which the results are required, the time available for the measurements, and the cost involved.

Mainly because of the many different purposes for which catalysts need to be tested, there is no one method suitable for all circumstances. Clearly a test which may be suitable for control purposes during catalyst manufacture may not be suitable when satisfactory process design information is required. Thus there is a large choice in the methods that may be used for catalyst testing, and in practice it is generally necessary to employ several different types of test (as shown in fig. 14) during the development of a new catalyst. Although the industrial catalyst user is interested in initial performance, he also wants information about stability, that is, on how well performance is maintained over a period of time. This means that the most appropriate catalyst testing methods are continuous constant pressure methods rather than batch constant volumes methods. Changes in catalyst performance with time can easily be seen with a continuous test because changes in conversion are reflected by changes in the concentrations of reactants and products in the exit stream.

There are two fundamental types of continuous test reactor: the integral tubular reactor, and the differential reactor. With an integral tubular reactor a reactant concentration profile will develop along the length of the catalyst bed, and the rate of reaction along the length of the converter will change. With a differential reactor the reactant concentration in contact with all parts of the catalyst bed will ideally be the same, and the rate of reaction will therefore be the same at all points in the converter. This is why the differential reactor is potentially the more suitable for the derivation of kinetic data, because the reaction rate can be calculated directly from the flow rate and the inlet and exit reactant concentrations. In the case of an integral reactor, an integration, which may be impossible to carry out algebraically, is often required.

Tubular reactors can be used for testing a catalyst, either under pore diffusion limited, or non-pore diffusion limited conditions. Pore-diffusion limitations can be avoided if the catalyst to be tested is crushed until the catalyst particles are of an appropriately small size. It has already been pointed out that it is difficult to extrapolate from powdered catalyst tests to the behaviour of full-size particles. Powdered catalyst tests, however, are very useful for rapidly screening and comparing the performance of a large number of catalyst formulations. Because of the absence of a diffusion limitation, this type of test is more sensitive to small changes in catalyst formulation than a test involving full size particles. In addition gas distribution problems are less likely to arise, but, unfortunately, such a test is also more sensitive to the effects of poor temperature control.

Integral reactors

Integral tubular reactors can be operated in three different ways: isothermally, adiabatically, or pseudo-isothermally (that is, neither isothermally nor adiabatically).

The simplest way to run an integral converter is isothermally, especially for the derivation of kinetic data, since this restricts the number of variables and facilitates integration. In practice, however, isothermal operation is rarely possible, particularly with reactions having a high heat of reaction, because of heat transfer limitations. These limitations are important because poor heat flow control, leading to small temperature differences in the bed, can have a very severe effect, since the rate of reaction is exponentially dependent on temperature.

With exothermic reactions, tubular adiabatic reactors are commonly employed in order to overcome this problem. The temperature-control system is arranged in such a way as to prevent heat flowing through the walls of the reactor. Consequently a temperature profile develops along the length of the reactor, the extent and shape of the latter depending on the heat of reaction, the specific heat of the fluid, and the reaction kinetics. Because of their low surface to volume ratio, full-scale plant converters normally operate adiabatically, and so small scale adiabatic converters are useful for life tests, or for the simulation of full scale operation. These converters can be operated either on the semi-technical scale, or as sidestream units. Adiabatic reactors are currently employed for the simulation of full scale plant conditions for such reactions as HT and LT carbon monoxide shift conversion, methanation, and ammonia synthesis. The control system for some of the semi-technical units is shown in fig. 17. The point has been made already that it is generally more difficult to derive a kinetic equation from integral data than from differential data. The degree of difficulty is increased when integral tubular reactors are operated adiabatically. Nevertheless, under these conditions it is possible to obtain a large amount of information from a relatively small number of experiments, and on occasions it is useful to use this type of data for the derivation of a design equation. In any case, a design equation obtained from differential data must be cross-checked against integral data, obtained both on the small scale and the full scale, before it can be used with any degree of confidence.

FIG. 17. The control system for some of the semi-technical converter units.

A useful but accurate short cut that is sometimes used for the comparison of activities of different catalysts in an integral converter, and in the absence of any kinetic information, is the measurement of space velocities for standard inlet and exit conditions. Under these circumstances space velocity is directly proportional to activity, and can thus be used to give an accurate comparison of different catalysts. Tubular reactors that are operated neither isothermally nor adiabatically are termed 'pseudo-isothermal'. Because the reaction conditions cannot usually be mathematically defined, this type of reactor is useless for the provision of a design equation. Nevertheless it is useful for control purposes, and to make quick comparisons between different catalysts. In general it is not possible to use the type of information obtained from this reactor for scale-up purposes. Reactors are generally operated in this way because of heat transfer limitations which occur with very endothermic or exothermic reactions. Isothermal operation is not possible in these cases because, in order to achieve a large surface–volume ratio, the diameter of the converter may be restricted below the practical limit. In such a case where heat transfer is limiting, a multi-tubular reactor may be the solution to the problem, and it can be convenient to simulate this on the small scale by a full-size tube. This avoids difficulties that would arise if scale-up were attempted from data obtained under neither isothermal nor adiabatic conditions. An excellent example of a highly endothermic reaction in this category is the steam-reforming reaction. The best means of testing reformer catalysts is in a full sized reformer tube, that is to say, on a semi-technical plant. This has the great advantage that it duplicates the full-scale operation, which is simply a multiplication of the number of tubes. Carefully designed small-scale tests, however, are of great value for an initial rapid screening of new catalyst formulations. Small integral tubular reactors are used, and heat is provided in such a way as to simulate the normal reactor temperature profile.

Differential reactors

The advantages of differential reactors for studies of reaction kinetics have already been discussed. Since the reaction rate is the same in all parts of the bed, changes in reaction rate caused by changes in each variable (temperature, pressure, concentrations of reactants, and products), can be studied separately. The procedure, however, can be time consuming, because a very large number of experimental results over a wide range of conditions may be needed before a design equation can be produced. Nevertheless, the exercise proves well worthwhile because very accurate results can be obtained.

Differential tubular reactors have an additional advantage over integral tubular reactors in that the uniformity of fluid properties throughout the bed ensures plug flow. It is necessary to add products as well as reactants to a differential system so that their effect on reaction rate can be ascertained. Disadvantages of the differential reactor are that high gas rates are needed to preserve differential conditions, and that side reactions, which are usually much slower than the main reaction, are often difficult to simulate, or can go undetected.

Three types of differential reactor are available. These are the single-pass tubular reactor, the recirculatory reactor, and the stirred tank reactor. Their names describe the way they work.

(1) *Single-pass tubular reactors*

The great advantage of this type of reactor is its simplicity, small size, and ease of operation. It can easily be maintained isothermally, because of the low conversions obtained, but it suffers from the disadvantage that very accurate composition analyses are required; since the reaction rate is calculated from the small differences between inlet and exit concentration. With modern methods of analysis this provides a sufficiently accurate method of catalyst comparison for a wide variety of reactions. Reasonably accurate kinetic data may also be measured.

(2) *Recirculatory reactors*

In those cases where the analytical errors associated with single pass differential conversion are too large, the use of a recirculation system may solve the problem (ref. 12). With this system, part of the exit gas is recycled and mixed with a small amount of fresh gas. After some initial changes the system will reach a steady state. The composition of the make-up gas and exit gas are then measured, together with the exit flow rate. Since the pass conversion

FIG. 18. Reactor used to study ammonia synthesis kinetics. A, reactor and heater; B, purification; C, pressure controller; D, flow controller; E, circulator.

depends on the recycle rate, which can be very large, the system can be made to approach very closely to true differential operation. With this type of reactor the rate of reaction is measured directly as follows (assuming no change of gas volume with conversion):

$$\text{reaction rate} = \frac{X_{in} - X_{out}}{V/F_{in}},$$

where X_{in} and X_{out} are the inlet and exit mole fractions of the reactant, V is the volume of the catalyst, and F is the flow rate of the reactant in and out of the system (moles/unit time).

Providing the unit is operating differentially, there is no change of reaction rate with circulation rate.

There are two main advantages with the recirculatory reactor. Firstly, the overall concentration difference between the feed gas and the exit gas can be very different, even though the converter is operating differentially, which decreases the importance of analytical errors. This depends, however, on the overall degree of conversion, and if this is allowed to become small the advantage of using a recirculatory system may be lost. Secondly, because the converter is operated differentially, temperature gradients within the converter will be small. Nevertheless, heat may have to be supplied or removed by the large volume of circulatory gas.

One disadvantage of this type of system which is sometimes overlooked is the potential for build-up of by-products or impurity concentrations. For this reason it is not used for reactions in which there is likely to be by-product formation (methanol synthesis, for example), but it has been used very successfully to study the kinetics of ammonia synthesis from nitrogen and hydrogen. A typical reactor which has been used to study the kinetics of ammonia synthesis is shown in fig. 18.

(3) *Continuous stirred tank reactors*

With this type of differential reactor, the reactants are fed continuously into a tank where they are perfectly mixed and brought into contact with the catalyst by a stirrer. In order to balance the inflow of reactants, gas which contains both the reactants and the products is continuously removed from the system.

Interest in these reactors for the study of the kinetics of catalytic gaseous reactions has recently been revived by Carberry (ref. 13) and by Brisk *et al.* (ref. 14). They have been used for a number of purposes within ICI, including a study of the kinetics of the low temperature shift reaction and the effect of poisons on the reaction. This type of reactor is particularly suitable for a differential study of poisoning reactions, since, in contrast to a tubular reactor, all the catalyst is exposed simultaneously to the same concentration of poison.

The simplifying assumption of perfect gas mixing is necessary if continuous stirred tank reactors are to be used for the derivation of kinetic data, so the occurrence of such mixing must be checked before carrying out experimental work. Provided that the fluid is not too viscous, and that stirring is adequate, the perfect mixing assumption can usually be justified. Experiments with residence times and gas distribution should be carried out under non-reacting conditions, so that the mixing pattern is established. If any deviations from ideal behaviour are discovered their causes should be traced and the situation corrected, because the experimental data resulting from a combination of the effects of both imperfect mixing *and* the chemical reaction, are usually too complex for accurate interpretation, and can therefore be misleading. Correctly used, continuously stirred tank reactors offer a versatile technique for the analysis of kinetic problems.

Strength and attrition testing

The general requirement of catalyst strength was noted in Chapter 2. Catalyst particles must be strong enough to withstand four different forms of stress. These are

(a) abrasion (during transit),
(b) impact (when loaded into the converter),
(c) internal stresses (resulting from phase changes—usually during reduction, or when initially brought on line), and
(d) external stressing (caused by pressure drop, catalyst weight and, possibly, thermal cycling).

It is clearly impossible to devise a test which will take account of all these factors. ICI therefore uses different tests for measuring the ability of catalyst particles to withstand mechanical stressing, abrasion and impact. For control purposes, a single strength measurement is generally adequate, because experience with production catalysts permits a subjective evaluation of the other properties. For new catalysts, however, measurements of each property must be made.

Crushing strength tests

The most commonly used mechanical tests for catalysts involve measuring the breaking strengths of individual pellets. This is done by increasing the load on a pellet until failure occurs, and the mean of at least ten such tests is taken as the crushing strength. For cylindrical pellets there are two such tests which differ simply in the way the load is applied.

In the 'vertical crushing test' the pellet is crushed with its flat ends in contact with two plane, rigid platens. For plant testing, the load is applied by a hand-wound screw, which forces the pellet against an upper platen attached to a piston in an oil filled cylinder coupled to a pressure gauge. The maximum load the pellet can withstand is observed visually on the gauge, and is recorded as the 'vertical crushing strength'. Pellets usually break by a 'double-cone' type of fracture, in which they disintegrate to small particles, leaving only a cone at each end. The mechanism of failure is complex, and is affected by the state of the platen (which must be kept clean and polished), and the length/diameter ratio of the pellet. For different sized pellets of the same material, with constant length/diameter ratio, the crushing load is proportional to the cross sectional area of the pellet. That is to say, the crushing stress is constant.

In the second crushing test for cylindrical pellets, the specimens are compressed along a diameter between two flat plates, or between a flat plate and a cylindrical bar at right angle to the pellet axis. Such loading produces a tensile stress perpendicular to the loaded diameter, and the pellet splits cleanly down the middle. The breaking load is directly proportional to the length and radius of the pellets, and to the tensile strength of the material. The relationship between the results of these two tests depends on pellet geometry, and on the type of material, but, typically, the VCS is ten times the HCS (horizontal crushing strength). Consequently, determination of the latter requires more sensitive testing equipment.

Other strength tests

The effect of impact on catalyst pellets is easily determined by dropping pellets on to a surface at least as hard as they are. The length of the drop needed to fracture the pellets, or a proportion of them, can be ascertained. Each catalyst formulation shows a fairly constant relationship (determined by its brittleness) between the drop needed for fracture and the crushing strength (see Chapter 2, pp. 26–28).

With some catalysts it is necessary to carry out strength tests under reaction conditions. This can be done in a special reactor fitted with a piston by means of which a load can be applied to the catalyst particles. The strength of the catalyst can be evaluated in terms of the percentage breakage or the proportion of fines produced.

The resistance of all types of catalyst to abrasion during transit can be measured by a tumbling test. In this test a constant volume of catalyst particles is subjected to a specified number of falls in a rotating cylindrical container. The tumbling loss is defined as the proportion (w/w per cent) which is reduced to a powder finer than 10 mesh.

CHAPTER 4

Desulphurisation

It is well known that transition metal catalysts used in a variety of chemical processes are prone to deactivation by certain chemical species which are able to donate electrons into the unfilled d-orbitals of the metal. A particular example of this is the deactivation of nickel-containing catalysts (such as the ICI steam reforming catalysts) by hydrogen sulphide or organic sulphur compounds. Therefore, to avoid deactivation, maximum limits are set on the sulphur content of the feed in contact with these catalysts. ICI define operating conditions such that desulphurisation of the hydrocarbon feedstocks prior to the steam reforming stage reduces the initial sulphur concentration down to the allowable sulphur level. The actual limit set depends on the conditions used in the steam reforming section.

Feedstocks for steam reforming

Hydrocarbon feedstocks used for the production of synthesis gas by steam reforming range from natural gas (essentially methane, CH_4, together with a few per cent of higher boiling hydrocarbons) and light petroleum gasoline (mainly butane C_4H_{10} with some butene and higher-boiling hydrocarbons), to light distillate fractions boiling in the range 40–170°C, which contain a range of hydrocarbon types (e.g. 65 per cent v/v paraffins, 25 per cent v/v naphthenes, 10 per cent v/v aromatics and 1 per cent v/v of olefines). In this case the average molecular weight approximates to 100 and the density is 0·68–0·72 g/cm³, which are similar to the molecular weight and density of heptane C_7H_{16}.

The types and concentrations of sulphur compounds present in these feedstocks depend on the boiling range and on the source. Natural gas contains mainly hydrogen sulphide plus low-boiling sulphides or mercaptans, such as methyl mercaptan and dimethyl sulphide. Light petroleum gasoline contains similar compounds of higher boiling point. Light distillates contain a large range of compounds of different types. In one typical light distillate, for example, 77 separate sulphur compounds have been characterised, including 36 mercaptans, 23 linear sulphides, 18 cyclic sulphides, and thiophene (ref. 15). Most distillates also contain some disulphides. The boiling range of the feedstocks, the total sulphur concentrations (typically in the range 50–600 ppm w/w S), and also the sulphur types present, are important in the design of the desulphurisation section.

Methods of desulphurisation

Natural gas

Sulphur removal from natural gas by absorption at ambient temperatures has been extensively used in North America. Activated charcoal or molecular sieves are mainly used as absorbents. Frequent regeneration of the absorbent is required, and two or more vessels must be available so that one is on line while the other is being regenerated. The efficiency of the absorbent systems depends both on the type of sulphur compounds and the concentration of higher hydrocarbons present in the natural gas. Low-boiling sulphur compounds are not strongly absorbed, and the presence of condensable hydrocarbons can rapidly saturate the absorbent. Thus when variations of these types occurs, the efficiency of desulphurisation is often unreliable. In this case the use of a guard vessel, containing an absorbent such as zinc oxide, is advisable.

If the natural gas contains mainly hydrogen sulphide and mercaptans, zinc oxide alone can be used, preferably at temperatures in the range 350–400°C. Where a range of organic sulphur compounds may be present, the preferred method is described in the following section.

Higher-boiling fractions

With higher-boiling feedstocks which may contain thiophenic compounds, desulphurisation is effected by hydrogenolysis of the organic sulphur compounds to hydrogen sulphide, which is then absorbed by zinc oxide. In the ICI process, the desulphurisation section consists of a single bed of cobalt molybdate sandwiched between two beds of zinc oxide. The first bed of zinc oxide has two functions, as both catalyst and absorbent, since the reactive sulphur compounds are decomposed to hydrocarbon and hydrogen sulphide, which is then absorbed. Thiophenic compounds and any remaining reactive compounds are hydrogenated to hydrogen sulphide over the cobalt molybdate, and this hydrogen sulphide is absorbed by the second zinc oxide bed. Thiophenic compounds are occasionally referred to as 'non-reactive sulphur'. In this context mercaptans, sulphides and disulphides are termed 'reactive sulphur'. Nickel molybdate is occasionally used as an alternative to cobalt molybdate, and iron oxide has been used as a cheap alternative for zinc oxide, although zinc oxide is a superior absorbent (see p. 51). The first zinc oxide bed is used to remove the reactive sulphur compounds since they, or the hydrogen sulphide formed by their hydrogenolysis, would reduce the rate of thiophene hydrogenolysis over the cobalt molybdate.

During use, the zinc oxide is converted to zinc sulphide and therefore the absorption function is progressively exhausted, and fresh zinc oxide must be charged. The lifetime of the zinc oxide obviously depends on the sulphur content of the feedstock. Therefore where high-sulphur feedstocks are in use, a prior desulphurisation stage is employed. This may consist of either an acid washing plant or a hydro-treating unit.

The acid washing plant consists of a sulphuric acid wash which removes the major part of the organic sulphur, typically from 400–500 ppm down to 30–100 ppm, depending on the conditions used. A major disadvantage of this system is the formation of acid sludge, bringing with it the problem of sludge disposal.

In the hydro-treating unit the sulphur compounds in the feedstock are converted to hydrogen sulphide in the presence of cobalt molybdate under conditions such that the final sulphur concentration is 5–10 ppm. The dissolved hydrogen sulphide is physically stripped from the distillate in a heated stripping column. In this case, since the sulphur compounds in the feedstock leaving the hydro-treating unit will generally be of the refractory type (which will not be affected by the first zinc oxide bed in a closed zinc oxide—cobalt molybdate–zinc oxide sandwich), an open sandwich will often be adequate.

The combination of systems used depends on the operator's estimate of the lifetime required for the zinc oxide catalyst, since this determines the running time of the plant between shutdowns for catalyst recharging.

ICI catalysts

ICI produces two types of desulphurisation catalyst, cobalt molybdate (catalyst 41) and zinc oxide (catalyst 32) which have been successfully used for many years in the desulphurisation of hydrocarbon feedstocks.

Cobalt molybdate

In common with most manufacturers, ICI produces both pelleted and extruded catalysts. Catalyst 41-4, which is extruded, has a lower bulk density than the pelleted catalyst 41-3, and thus, for a given catalyst volume, a lower weight of 41-4 is charged. However, because of the higher surface area and porosity of 41-4, the catalysts have an equivalent activity under reaction conditions, and the same space velocities are used for a given degree of sulphur removal.

Catalyst 41-3 has been used extensively with the ICI hydrodesulphurisation sandwich system, and has given service for periods up to five years without the need for regeneration, no increase in pressure drop through the plant being observed, indicating that strength remains good. Catalyst 41-4 has given similar service since its introduction three years ago.

The susceptibility to carbon formation, for different catalysts, can be ascertained by comparative tests under carbon-forming conditions. A typical rate of carbon formation for most commercial catalysts, under these conditions (per kilo of catalyst), is 1·2 g carbon hr^{-1}. Catalyst 41-4 is a modified formulation specially developed to minimise carbon formation and consequent deactivation during use. This decreases the need for catalyst regeneration in hydro-treaters and hydro-desulphurisers. However, some deposition of carbon takes place with all catalysts of this type (it is particularly marked when low-pressure–high-temperature combinations are employed). The comparative rate of carbon formation on catalyst 41-4 is 0·33 g h^{-1}.

The catalyst essentially consists of a mixture of cobalt oxide and molybdenum oxide supported on high surface area alumina. Cobalt oxide, molybdenum oxide, and other transition metal oxides taken individually, are active for the hydrogenolysis of organic sulphur compounds, but maximum activity is shown with the above mixture. There is some speculation concerning the active phase, or phases, formed from this mixture which are responsible for the enhanced activity. The fresh catalyst consists of Al_2O_3, $CoAl_2O_4$, CoO, MoO_3, $CoMoO_4$, and a complex cobalt-molybdenum oxide (ref. 16). The catalyst, normally charged in the oxide state, shows activity for the hydrogenolysis reaction, but optimum activity is not reached until the catalyst becomes sulphided. Under desulphurisation conditions the catalyst consists of Al_2O_3, $CoAl_2O_4$ (inactive) Co_9S_8 (slightly active) MoS_2 (moderately active) plus some MoO_2. The 'active' catalyst has been described as MoS_2 promoted with non-reducible cobalt (ref. 16). It seems possible that the function of the cobalt oxide is to minimise sintering of the molybdenum sulphide crystallites by keeping them apart (see pp. 23–25) and to increase activity by increasing the effective surface area of the molybdenum sulphide phase.

Under typical conditions the catalyst absorbs 1–3 per cent sulphur before equilibrium is reached. The actual sulphur content is determined by the ratio of partial pressures of hydrogen sulphide and hydrogen in contact with the catalyst. Thus an increase in the sulphur content of the feed, or a decrease of the hydrogen pressure, can cause the catalyst to gain sulphur as the ratio is increased. And it can lose sulphur by the reverse of these changes. The sulphur content of the mixed oxide/sulphide catalyst can be estimated from the hydrogen reduction data of the individual sulphides.

The term 'cobalt molybdate' is used generally to describe the mixture of cobalt oxide and molybdenum oxide supported on alumina, as described above.

Zinc oxide

ICI zinc oxide catalyst (32–4) is in the form of $\frac{1}{8}-\frac{3}{16}$ in diameter granules. Catalysts of this type have been in general manufacture for over twenty years. Straight zinc oxide catalysts tend to lack porosity and hence the rate of sulphur pick-up is limited (though total sulphur pick-up may still be high). ICI catalyst 32–1 was a catalyst of this type. The zinc oxide in catalyst 32–4 is made accessible by a modified pore-structure. 32–4 has found general use in the removal of organic sulphur compounds and hydrogen sulphide from a variety of gas streams, particular examples being in the ICI desulphurisation sandwich, and as a sulphur guard for low-temperature shift catalysts.

In use, the catalyst is converted to zinc sulphide, and while zinc sulphide maintains an activity for the decomposition of organic sulphur compounds, the absorption of hydrogen sulphide ceases when the oxide becomes fully converted. At this stage the catalyst must be replaced. The maximum usage of this catalyst as an absorbent depends on the following factors:
 (a) The amount of zinc oxide present per unit volume. This depends on a combination of the zinc content and the bulk density.
 (b) The availability of the zinc oxide in a given catalyst volume, which depends on the pore volume and effective surface area.
 (c) The resistance to breakdown during transport, charging, and use.

These factors are related (see fig. 5). Increasing the pore volume, for example, can increase the availability of the zinc oxide, but also causes a decrease in bulk density and strength, and therefore a careful optimisation must be made.

If decomposition and absorption of sulphur compounds takes place only on the outside surface of the granules, the catalyst activity is low. For high activity the sulphur-bearing gas steam must enter the granule interior through the pores. Migration of the sulphide ions into the oxide crystallites, and migration of the oxide ions out, must also take place in order to utilise the zinc oxide present fully. Thus a finite time is required for reaction to occur, and the rate of reaction increases with increasing temperature. The overall rate of decomposition and absorption is dependent on the type and concentration of the sulphur compound, the gaseous space velocity, and reaction temperature. Under perfect

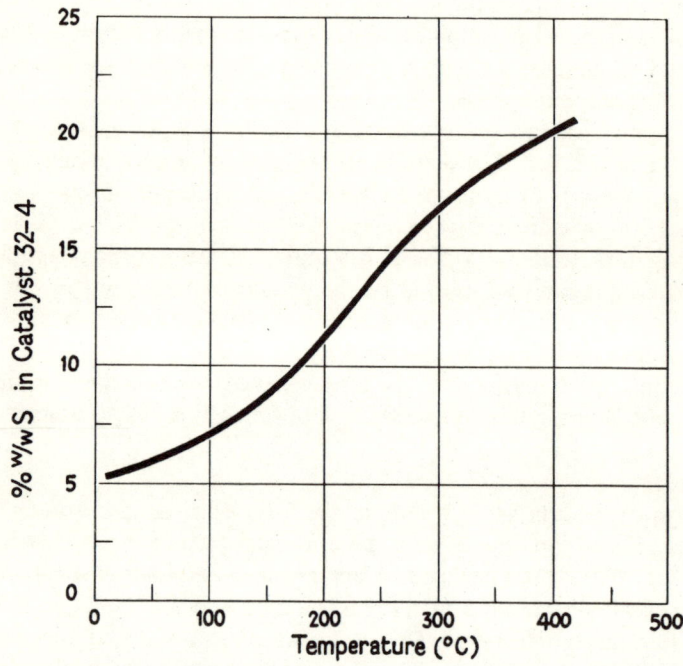

FIG. 19. Variation with temperature of sulphur content of 32-4 catalyst at breakthrough. Gaseous space velocity, 700 h^{-1}, inlet H_2S, 50 ppm, exit H_2S, less than 1 ppm.

conditions, sulphur would be absorbed by each finite section of the bed until it was fully sulphided, and the sulphur absorption front would be sharp.

In practice, the reaction temperature is not higher than 400°C in order to minimise cracking of the feedstock, and the gaseous space velocity is determined by the throughput and the lifetime required. Under these conditions the sulphur absorption front is diffuse, and although the catalyst at the bed inlet becomes fully sulphided, breakthrough of sulphur occurs before the catalyst at the exit of the bed is fully saturated. Sulphur breakthrough is said to occur when the sulphur concentration measured at the bed exit exceeds a certain value—for example 0·5 ppm. The catalyst activity may therefore be defined in terms of the sharpness of the sulphur front, which is measured by the overall sulphur content of the catalyst bed at breakthrough. The sharpness of the front, being related to the rate of reaction between sulphur and the catalyst, varies with the types of sulphur compounds in the feed gas. H_2S gives a relatively sharp front, as do mercaptans, but less reactive sulphur compounds produce more diffuse fronts.

The performance of a charge of zinc oxide catalyst can be assessed only by determining the average sulphur content of the bed (the inlet always approaches theoretical and the exit contains very little sulphur). It is therefore convenient to measure performance in terms of the spent catalyst. Catalyst 32–4, operated at 370°C and a space velocity of 400 h^{-1} until breakthrough is noted, normally achieves an average sulphur absorption in the range 22–24 per cent w/w S (this is equivalent to 24·4–26·7 kg sulphur per 100 kg fresh catalyst). From a design standpoint, under the above conditions, a sulphur content corresponding to an average of 18 per cent w/w S (equivalent to 20 kg sulphur per 100 kg charged) is a useful level to consider because ICI guarantee this to be exceeded. The sulphur content decreases with decreasing temperature or increasing space velocity. The variation with temperature is shown in fig. 19.

Alternative catalyst absorbents

In general, other manufacturers have had less incentive than ICI to produce zinc oxide catalysts which balance activity and absorption capacity, density,

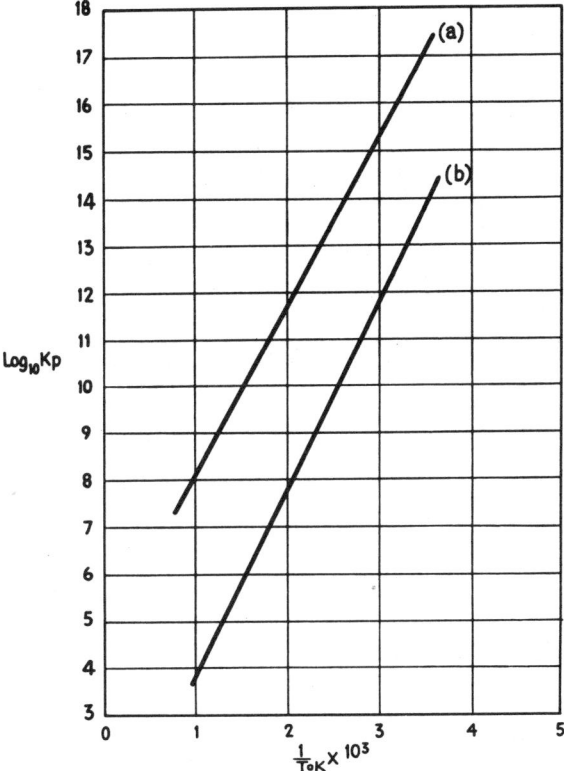

FIG. 20. H_2S evolution in presence of water vapour. Equilibrium data for
(a) $Fe_3O_4 + 3H_2S + H_2 \rightleftharpoons 3FeS + 4H_2O$,
(b) $ZnO + H_2S \rightleftharpoons ZnS + H_2O$.

Temperature (°C)	(a) Equilibrium H$_2$S (ppm v/v)					(b) Equilibrium H$_2$S (ppm v/v)				
	200	250	300	370	400	200	250	300	370	400
H$_2$O (per cent)										
3.3	1.85	3.1	5.1	8.9	10.8	2.6×10^{-4}	1.7×10^{-2}	0.7×10^{-2}	4.2×10^{-2}	6.5×10^{-2}
1.7	7×10^{-1}	1.35	2.02	3.51	4.23	1.3×10^{-4}	0.9×10^{-3}	0.3×10^{-2}	2.1×10^{-2}	3.2×10^{-2}
0.33	0.85×10^{-2}	1.57×10^{-1}	2.01×10^{-1}	4.1×10^{-1}	4.9×10^{-1}	2.6×10^{-5}	1.7×10^{-3}	0.7×10^{-3}	4.0×10^{-3}	6.5×10^{-3}
0.17	1.57×10^{-3}	0.62×10^{-1}	0.93×10^{-1}	1.64×10^{-1}	1.96×10^{-1}	1.3×10^{-5}	0.9×10^{-4}	0.3×10^{-3}	2×10^{-3}	3.3×10^{-3}

FIG. 21. The effect of water vapour and temperature on equilibrium H$_2$S concentration. Reactions
(a) $Fe_3O_4 + 3H_2S + H_2 \rightleftharpoons 3FeS + 4H_2O$,
(b) $ZnO + H_2S \rightleftharpoons ZnS + H_2O$.
Gas contains 20% hydrogen.

and strength. The main alternative desulphurisation catalyst is iron oxide, which is generally available as extrusions. The advantages of iron oxide lie in its low cost and, theoretically at any rate, in its potential for regeneration. The partial pressure of hydrogen sulphide in the gas stream emerging from an iron oxide bed is markedly affected by operating conditions, so much more stringent control of conditions is required with iron oxide than with zinc oxide. The difference in the performance of the two absorbents is related to the effect of water vapour on the sulphur absorption equilibria (sometimes, also, to the effect of hydrogen on zinc sulphide and iron sulphide).

The reactions which take place during use of iron oxide or zinc oxide as absorbents for hydrogen sulphide can be compared. Fresh iron oxide consists essentially of Fe_2O_3, which is converted to Fe_3O_4 in the presence of hydrogen above approximately 175°C. (In normal use, in the temperature range 340–400°C, Fe_3O_4 is the absorbent.)

$$3 Fe_2O_3 + H_2 \rightleftharpoons 2 Fe_3O_4 + H_2O$$

$$Fe_3O_4 + H_2 + 3 H_2S \rightleftharpoons 3 FeS + 4H_2O$$

For zinc oxide, the corresponding reaction is

$$ZnO + H_2S \rightleftharpoons ZnS + H_2O$$

Thus, both equilibria are affected by the presence of water vapour, which may be initially present in the hydrogen or hydrocarbon feedstock, or which may be formed, for example, by methanation of carbon oxides on cobalt molybdate. The effect of water partial pressure on the equilibrium hydrogen sulphide partial pressure in contact with Fe_3O_4 or ZnO over the range 200–400°C can be estimated from fig. 20. Fig. 20 has been calculated for a plant operating with a hydrogen : hydrocarbon mole ratio of 0·25. Fig. 21 illustrates how the equilibrium values of H_2S partial pressure vary with both temperature and the concentration of water vapour.

With low water partial pressure (less than 0·2 per cent H_2O) the reaction is probably kinetically controlled rather than equilibrium controlled in both cases. Although the equilibrium hydrogen sulphide partial pressure is increased in both cases as the proportion of water is increased, with zinc oxide the hydrogen sulphide partial pressure remains very low over a wide range of water concentrations, and the reaction remains kinetically controlled over the range quoted in fig. 21. With iron oxide, however, the reaction becomes equilibrium controlled, and the partial pressure of hydrogen sulphide is sufficiently high to decrease the activity of the steam-reforming catalyst.

A further potential disadvantage of iron oxide (compared with zinc oxide) is associated with the relative ease of reduction of the sulphides, according to the equation:

$$MS + H_2 \rightleftharpoons M + H_2S$$

The variations of equilibrium constants with temperature for the two reactions are shown in fig. 22 and the critical ratios of $H_2S:H_2$ required to maintain a sulphided absorbent are shown in fig. 23. At 400°C, for FeS the equilibrium concentration of hydrogen sulphide in hydrogen is 19 ppm v/v.

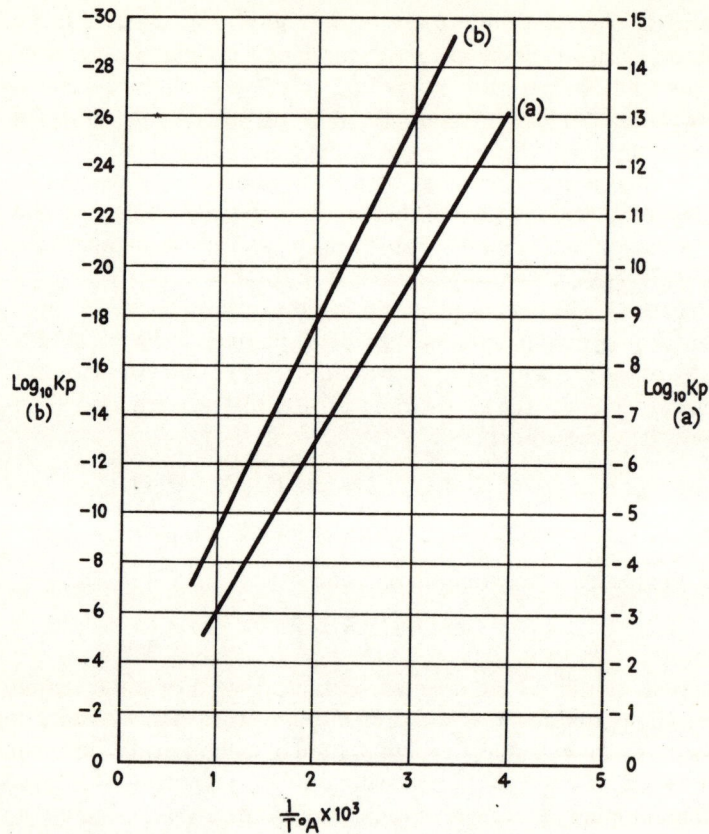

FIG. 22. H_2S evolution in presence of hydrogen. Equilibrium data for (a) $FeS + H_2 \rightleftharpoons Fe + H_2S$, (b) $ZnS + H_2 \rightleftharpoons Zn + H_2S$.

Reactions: (a) $FeS + H_2 \rightleftharpoons Fe + H_2S$
(b) $ZnS + H_2 \rightleftharpoons Zn + H_2S$
Gas approximately 100% H_2.

Temperature (°C)		200	300	350	370	400
Equilibrium concentration of H_2S in H_2 (ppm v/v)	(a)	1.3×10^{-1}	2.5	7.8	10.7	19.1
	(b)	3.2×10^{-13}	4×10^{-10}	5×10^{-9}	1×10^{-8}	2×10^{-7}

FIG. 23. The effect of hydrogen and temperature on equilibrium H_2S concentration.

This is 10^8 times greater than the equivalent figure over ZnS. This problem is magnified if the flow of hydrocarbon feedstock is temporarily halted, e.g. through a plant malfunction, and the flow of hydrogen gas is maintained. With an iron oxide absorbent, the hydrogen sulphide concentration produced is large and it may be necessary to box up the desulphurisation vessel under an inert atmosphere in order to minimise the quantity of sulphur passing into the steam-reforming section, whereas with the zinc oxide absorbent the amount of hydrogen sulphide is completely insignificant and no precautions are necessary.

Some operators who use iron oxide also use a final guard bed of zinc oxide to minimise the chance of hydrogen sulphide passing from the desulphurisation section. Since for optimum steam-reforming efficiency, however, the sulphur concentration must consistently be maintained at a low figure, the dependability of the sulphur-removal section is of prime importance, and many users (including ICI), prefer to use zinc oxide alone as the sulphur absorbent.

Some principles of desulphurisation

The desulphurisation of hydrocarbon feedstocks in a hydro-treating unit depends on the hydrogenolysis of organic sulphur compounds to hydrogen sulphide over a cobalt molybdate catalyst, and sulphur removal with an ICI sandwich desulphuriser depends on a combination of this reaction with the decomposition and absorption of organic sulphur compounds and hydrogen sulphide on zinc oxide. An understanding of the design and operation of such systems requires a knowledge of the relative rates of reaction of different types of sulphur compounds, since this determines the conditions required for a given degree of sulphur removal. In some cases the thermal decomposition of certain types of sulphur compounds can contribute to the formation of hydrogen sulphide.

Thermal dissociation of organic sulphur compounds

The thermal decomposition of primary and secondary mercaptans take place at approximately 200–250°C, and tertiary mercaptans decompose at lower temperatures. The products are mainly olefines and hydrogen sulphide, although the reactions are free radical in nature, and complex products, including polymers, are also formed. Aromatic thiols are generally more stable. Aliphatic disulphides decompose at similar temperatures to mercaptans, giving a mixture of products which usually contains mercaptan, sulphide and hydrogen sulphide (ref. 17). Aromatic disulphides are more stable and tend to form sulphide and elemental sulphur at approximately 300°C.

Straight-chained and cyclic sulphides are generally stable up to 400°C, when the formation of hydrogen sulphide and olefines occurs through complex intermediates. Thiophenes are thermally stable at 470–500°C. The temperatures at which a range of sulphur compounds start to decompose, under the same conditions, is given in fig. 24 (ref. 18).

There is therefore a marked distinction between the thermal stability of mercaptans and disulphides which thermally decompose at 150–250°C, and sulphides and thiophenes which are stable at 400°C.

Some thermal decomposition can therefore take place in the preheater/vaporiser section of the desulphurisation unit prior to contact with the catalyst, with the production of hydrogen sulphide. The degree of decomposition depends not only on the types of sulphur compounds present in the feedstock, but also on the residence time in the preheater section. This reaction does not alter the

Compound	Temperature at which decomposition commences (°C)
n-C_4H_9SH	150
i-C_4H_9SH	225–250
$C_6H_{11}SH$	200
C_6H_5SH	200
$(C_6H_5)_2S$	450
$(C_2H_5)_2S$	400
$C_6H_5S\ C_6H_{11}$	350
thiophene	stable at 500
2,5-dimethylthiophene	475

FIG. 24. Thermal decomposition of a range of sulphur compounds.

total sulphur concentration in the stream but rather converts one type of sulphur compound into another (e.g. mercaptan into hydrogen sulphide).

Reactions on zinc oxide

Zinc oxide absorbs any hydrogen sulphide present in the feed, including that formed by thermal decomposition in the preheater. Further, the decomposition of organic sulphur compounds to hydrocarbon and hydrogen sulphide is also catalysed by zinc oxide. At high temperatures the removal of mercaptans and disulphides may be assisted by the thermal reaction (often by conversion to hydrogen sulphide in the preheater), but the removal of linear and cyclic sulphides is undoubtedly catalytic since these compounds can react on the catalyst at temperatures at which they are thermally stable. Thiophene does not decompose easily, and under ICI sandwich conditions the degree of removal on zinc oxide is negligible. Thus, thiophenic compounds are sometimes referred to as 'non-reactive sulphur', and other sulphur types as 'reactive sulphur'.

The rate of decomposition of mercaptans, disulphides, and sulphides is directly proportional to the sulphur-compound partial pressure, and almost independent of the hydrocarbon partial pressure. For thiophene the rate of decomposition is markedly affected by hydrocarbon partial pressure, the order of reaction being -0.5 with respect to the partial pressure of heptane. First-order rate constants for dimethyl disulphide and thiophene decomposition are $1 \times 10^{-1}\ h^{-1}$ (200°C) and $2 \times 10^{-4}\ h^{-1}$ (370°C) respectively, and therefore the low rate of thiophene decomposition is due to an inherent low activity coincident with the marked inhibiting effect of hydrocarbon.

Since sulphur removal is basically due to a catalytic decomposition reaction, rather than to a hydrogenolysis reaction, the main effect of increasing hydrogen pressure is to decrease the hydrocarbon partial-pressure. This increases the rate of reaction, and there is also a small contribution from the hydrogenolysis reaction. Some hydrogenation of the hydrocarbon fragments produced on the catalyst surface does take place, and the fragments are desorbed. The hydrogenating species are probably hydrogen atoms produced by the further decomposition of hydrocarbon fragments, although some adsorption of hydrogen from the gas phase may occur.

During use, zinc oxide is converted to zinc sulphide, which is still active for the decomposition of organic sulphur compounds. It is, however, usual to recharge the catalyst when breakthrough of hydrogen sulphide occurs, since the main purpose of the zinc oxide is to remove sulphur by absorption.

The hydrogenolysis of organic sulphur compounds

The hydrogenolysis of organic sulphur compounds in the presence of cobalt molybdate takes place over a large range of temperatures and pressures. Where the sulphur content of a feedstock must be consistently reduced to a specified low level, however, the reaction conditions must be carefully defined.

Basically, the hydrogenolysis reaction is carried out to convert organic sulphur compounds into hydrogen sulphide, which is then easily removed, using either physical stripping, as in a hydro-treater, or chemical absorption, as for example on to ZnO, as in a sandwich system. Some typical hydrogenolysis reactions for a range of sulphur compounds are as follows:

$$C_2H_5SH + H_2 \rightleftharpoons C_2H_6 + H_2S$$
$$C_6H_5SH + H_2 \rightleftharpoons C_6H_6 + H_2S$$
$$CH_3SC_2H_5 + 2H_2 \rightleftharpoons CH_4 + C_2H_6 + H_2S$$
$$C_2H_5SSC_2H_5 + 3H_2 \rightleftharpoons 2C_2H_6 + 2H_2S$$
$$C_4H_8S \text{ (tetrahydrothiophene)} + 2H_2 \rightleftharpoons C_4H_{10} + H_2S$$
$$C_4H_4S \text{ (thiophene)} + 4H_2 \rightleftharpoons C_4H_{10} + H_2S$$

The reactions are exothermic, and the heat produced depends partly on the sulphur type (which determines the number of moles of hydrogen take-up), and partly on the actual compound. Some typical values are given in fig. 25 (ref. 19). With the sulphur concentrations found in hydrocarbon fractions used in steam reforming (typically less than 1000 ppm) the heat rise is insignificant.

Reaction	$\Delta H_{(700 \text{ K})}$ (kcal mole^{-1})
$C_2H_5SH + H_2 \rightarrow C_2H_6 + H_2S$	−16·77
$C_2H_5SC_2H_5 + 2H_2 \rightarrow 2C_2H_6 + H_2S$	−27·99
C_4H_8S (tetrahydrothiophene) $+ 2H_2 \rightarrow$ n-$C_4H_{10} + H_2S$	−28·73
C_4H_4S (thiophene) $+ 4H_2 \rightarrow$ n-$C_4H_{10} + H_2S$	−66·98

FIG. 25. Heats of hydrogenation of some organic sulphur compounds.

The equilibrium constants for the hydrogenolysis of organic sulphur compounds are large and positive even at temperatures as great as 500°C (see fig. 26). Since the rate of hydrogenolysis increases with increasing temperature, the operating temperature is generally in the range 340–400°C. The reactions are typically carried out with hydrogen: hydrocarbon mole ratios of 0·25 to 0·5. That is, with hydrogen: organic sulphur mole ratios of approximately 250:1 to 1000:1. Under these conditions the reactions are essentially quantitative.

For example, at 370°C with an initial thiophene concentration of 500 ppm w/v S and a hydrogen:hydrocarbon mole ratio of 0·5, the equilibrium thiophene concentration is approximately 2×10^{-4} ppm w/v S in liquid naphtha.

Although there is relatively little information of a quantitative nature avail-

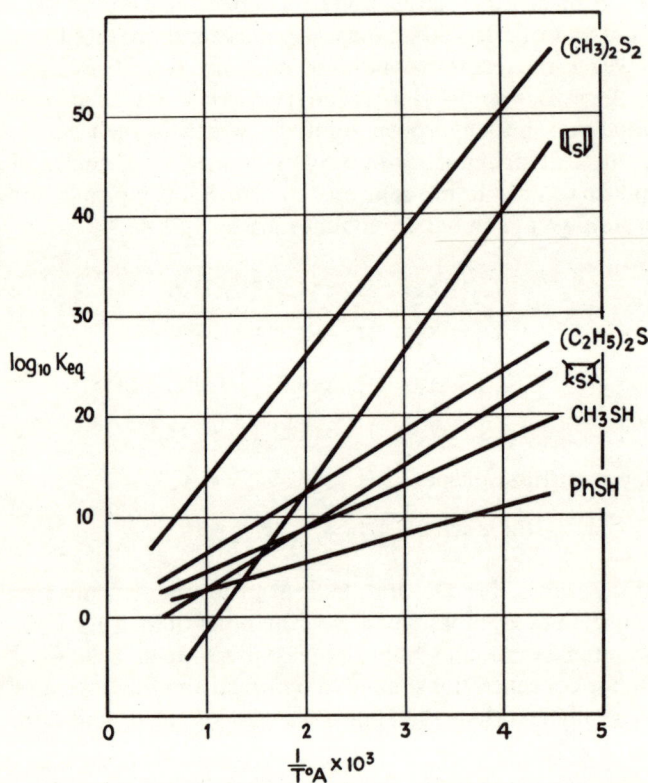

FIG. 26. Variation of equilibrium constants for the hydrogenolysis of some organic sulphur compounds, with temperature.

able in the literature concerning the ease of hydrogenolysis of different sulphur compounds, the following general points can be made:
(a) The hydrogenolysis of thiophene is less easy than the hydrogenolysis of sulphides, mercaptans, and disulphides.
(b) The ease of hydrogenolysis of the C_4 hydrocarbons increases along the series thiophene, 1,2-dihydrothiophene, tetrahydrothiophene, n-butyl-mercaptan (ref. 20).
(c) For a given type of sulphur compound, the rate of hydrogenolysis increases with increasing molecular weight.

It is obviously important to know how the rate of desulphurisation of a given hydrocarbon fraction depends on the rate of hydrogenolysis of the sulphur types present. In order to understand this in a more quantitative way, the rates of hydrogenolysis of five types of sulphur compound have been determined using ICI cobalt molybdate at atmospheric pressure and 370°C (ref. 21). The feedstock comprised n-heptane containing either dimethyl disulphide, diethyl sulphide, phenyl mercaptan, tetrahydrothiophene, or thiophene at concentrations of 100–500 ppm w/v S. The variation of the degree of heptane desulphurisa-

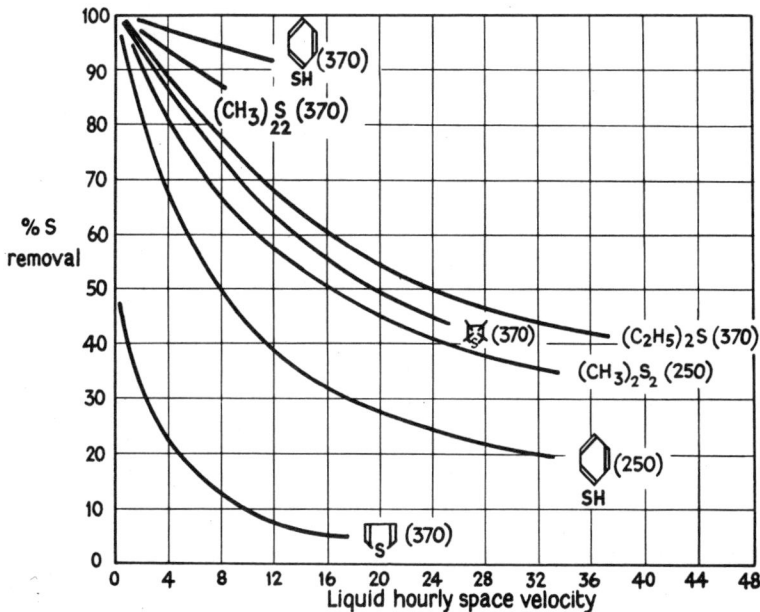

FIG. 27. Heptane desulphurisation: Catalyst 41–4. (Atmospheric pressure, 250°C, 370°C, $H_2:Hc = 0.25$.)

tion over a wide range of liquid space velocities using these compounds is given in fig. 27.

A marked difference is observed between the rate of hydrogenolysis of thiophene and the other compounds taken. Of particular importance, because of its potential use as a stenching agent, is the observation that tetrahydrothiophene, which has a cyclic structure like thiophene but is fully saturated, reacts at a rate similar to the linear sulphide, diethyl sulphide, rather than at a rate similar to thiophene. The rate of desulphurisation of a feedstock containing a mixture of sulphur compounds will be mainly controlled by the rate of hydrogenolysis of the most refractory compounds, which clearly are the thiophenic compounds.

Relatively little work has been published describing the kinetics of desulphurisation of specific sulphur compounds in the presence of cobalt molybdate. Published work has mainly involved the hydrogenolysis of a range of sulphur compounds, over various transition metal sulphide catalysts, in the absence of hydrocarbon. Orders of reaction range from 1 to 0 with respect to the sulphur compound, and 0.4 to 0.9 with respect to hydrogen. The activation energies are in the range 20–30 kcal mole^{-1}. Orders of 1 with respect to both sulphur compound and hydrogen are reported for the desulphurisation of n-heptane containing ethyl mercaptan, thiophene, or carbon disulphide in the presence of cobalt molybdate (refs. 22, 23, 24) and the activation energies are approximately 5 kcal mole^{-1}. Orders of 1 with respect to the sulphur concentration have also been observed during heavy naphtha desulphurisation over cobalt molybdate at elevated pressure (refs. 25, 26).

In the work carried out by ICI with the system described on p.47 the rate of heptane desulphurisation is also close to first order with respect to the sulphur compound concentration. Some first-order rate constants are given in fig. 28.

Temperature (°C)	K (g mol h^{-1} g catalyst^{-1} atm$^{-\frac{1}{2}}$)				
	RSSR	RSH	RSR	thiolane	thiophene
200	4.2×10^{-2}	1.8×10^{-2}	0.9×10^{-2}	3.7×10^{-3}	5.0×10^{-4}
250	10^{-1}	5.7×10^{-2}	3.1×10^{-2}	1.8×10^{-2}	1.7×10^{-3}
300	1.7×10^{-1}	1.2×10^{-1}	6.3×10^{-2}	4.8×10^{-2}	3.6×10^{-3}
350	2.6×10^{-1}	2.1×10^{-1}	1.1×10^{-1}	1.1×10^{-1}	7.1×10^{-3}
400	3.7×10^{-1}	3.6×10^{-1}	1.4×10^{-1}	2.2×10^{-1}	1.2×10^{-2}

FIG. 28. **First-order rate constants for heptane desulphurisation.**

In all cases, the rate of desulphurisation increases with increasing hydrogen partial pressure. For the cyclic compounds thiophene and tetrahydrothiophene the order of reaction is 0.5, and for diethyl sulphide, dimethyl disulphide, or phenyl mercaptan, 0.25. The rate of desulphurisation is also affected by hydrocarbon partial pressure, and the order of reaction with respect to heptane is -0.5 and -0.2 to -0.3 respectively for the two groups of compounds. This indicates that the hydrocarbon feedstock inhibits the hydrogenolysis reaction, probably by strong adsorption on the catalyst surface, thereby reducing the fraction of surface available for hydrogen and sulphur compound adsorption. An increase in the hydrogen:hydrocarbon mole ratio therefore increases the rate of desulphurisation by increasing the hydrogen partial pressure, and also by reducing the hydrocarbon partial pressure. The inhibiting effect decreases with decreasing molecular weight of the hydrocarbon. For example, with methane feedstock containing tetrahydrothiophene, the order of reaction with respect to methane partial pressure is -0.05.

The desulphurisation rate is increased as the total pressure is increased, with a measured pressure coefficient of 0.5 to 0.6. Also, activation energies observed in this work are in the range 10–13 kcal mole^{-1}, a magnitude approximately half those quoted in the literature for similar compounds (ref. 27). These factors are in agreement with the generally expressed view that desulphurisation reactions are pore-diffusion-controlled under plant operating conditions.

The required liquid space-velocity at elevated pressures, for the desulphurisation of naphtha containing either thiophene or tetrahydrothiophene, can be estimated from the rate constants measured at atmospheric pressure using the relationship

$$\text{rate of desulphurisation} = k \frac{P_S}{P^{\frac{1}{2}}} \left[\frac{P_{H_2}}{P_{Hc}} \right]^n,$$

where $n = 0.5$ for thiophene and tetrahydrothiophene, or $n = 0.25$ for mercaptan sulphide, or disulphide. For example, for a feedstock of average molecular weight 100 and density 0.7 g/cm^3 containing 50 ppm w/v S as thiophene, the estimated liquid hourly space velocity required to reduce the sulphur concentration to 0.5 ppm w/v S is 0.68, when operating with a hydrogen:hydrocarbon mole ratio of 0.25 at 30 atm pressure and a temperature of 370°C. Under the same conditions, 50 ppm w/v S as tetrahydrothiophene will be reduced to 0.5 ppm w/v S at a liquid hourly space velocity of 12 (a catalyst density of 1 g/cm^3 was taken for both cases). In a feedstock containing a range of sulphur compounds the presence of the more reactive sulphur reduces the rate of hydrogenolysis of thiophene, and the liquid space velocity required for a given degree of conversion must be decreased. This is the situation in a hydro-treating unit, like an open sandwich system (a bed of cobalt molybdate followed by a bed of zinc oxide). With a closed zinc oxide/cobalt molybdate sandwich the reactive compounds are removed by the first bed of zinc oxide, and the space velocity with respect to the cobalt molybdate bed can be designed on the rate constant for thiophene hydrogenolysis, unaffected by the presence of other compounds.

The actual design of conditions for the desulphurisation of a given feedstock is based on the considerations outlined above, and also on observations made over a range of temperatures and pressures on both semi-technical and plant converters.

Deactivation of cobalt molybdate

Three types of deactivation of cobalt molybdate catalysts may be defined. Temporary deactivation is observed in the presence of certain gaseous species, and removal of the gaseous species results in a return to the original activity. Carbon formation may be termed 'semi-permanent' deactivation since the initial activity can be regained by a regeneration process. Permanent deactivation can take place by loss of surface area or molybdenum during regeneration, or by the presence of certain components, for example, arsenic, which form compounds which are inactive for the hydrogenolysis reaction. In the last case a fresh charge of catalyst is required.

The effect of gas impurities

Hydrogen gas used in the hydrodesulphurisation of hydrocarbon feedstock invariably contains other gaseous components, such as N_2, NH_3, H_2S, CO, CO_2, and certain limits are set on their concentrations.

Sulphided cobalt molybdate causes hydrogenolysis of sulphur compounds by reaction principally on two sets of acid sites, one strongly acidic and the other more weakly acidic (ref. 27). The strong acid sites are sufficiently electrophilic to interact with olefines and enable them to be hydrogenated. Hydrogen sulphide and sulphur compounds are strongly absorbed on these sites which are also active for some hydrogenolysis of the sulphur compounds. The weaker acid sites, which are responsible for the major part of the hydrogenolysis reaction are poisoned by strong bases, like ammonia.

Thus, ammonia causes a reduction in the rate of desulphurisation by being adsorbed on the acid sites which are necessary for sulphur compound adsorption prior to reaction. The degree of deactivation is proportional to the ammonia partial pressure, and therefore for a specified catalyst volume the allowable ammonia concentration in hydrogenation gas is defined. For general ICI sandwich desulphurisation conditions, the level is 100 ppm v/v, although higher levels can be tolerated by adjusting the operating liquid space velocity.

High levels of N_2 can be tolerated, since nitrogen is only weakly adsorbed, and the rate of production of ammonia from nitrogen and hydrogen is small under desulphurisation conditions. When gas recycle systems are in operation, the ammonia level could build up slowly, and an ammonia scrubbing system might be necessary.

Hydrogen sulphide, in excess of that required to sulphide and activate the catalyst, also causes a reduction in the rate of hydrogenolysis of other sulphur compounds. With a closed ICI desulphurisation sandwich, the hydrogen sulphide is absorbed in the first bed of zinc oxide, and therefore has no effect on the reaction on the cobalt molybdate. In hydro-treating systems the concentration of hydrogen sulphide in the hydrogenating gas must be taken into account in the design calculations for determining the operating space velocity.

A limit is also set on the allowable concentrations of carbon monoxide and carbon dioxide present in hydrogenating gas. This is because of two factors. First, carbon oxides absorb on the acid sites and can therefore reduce the rate of desulphurisation, although the effect is less marked than with ammonia. In the second place, the carbon oxides undergo methanation on cobalt molybdate at temperatures above 300°C, resulting in the production of heat (see Chapter 6), and also a reduction in the hydrogen partial pressure, which decreases the desulphurisation rate (pp. 60–61). Allowance can be made for the temperature rise in the presence of a large excess of hydrocarbon, but any plant malfunction which results in the loss of hydrocarbon feed causes an appreciable temperature rise, possibly causing damage to the catalyst. The carbon oxides concentration is therefore generally set at a total of 5 per cent v/v of the hydrogen in the gas, including 3·5 per cent v/v CO and 1·5 per cent v/v CO_2.

Carbon formation

Cobalt molybdate catalysts gradually accumulate a carbonaceous deposit during use, which results in a progressive decrease in activity. The 'carbon' or 'coke' comprises high molecular weight polymeric compounds, and also, possibly, some free carbon. Precursors of the coke are carbon fragments which are probably formed by a mixture of catalytic cracking (which is assisted by low hydrogen pressures) and thermal cracking (which is assisted by high temperatures). For this reason recommendations are made concerning the minimum hydrogen partial pressures and the maximum temperatures which should be used in conjunction with these catalysts.

A 'coked' cobalt molybdate can be regenerated by using controlled oxidation to burn off the carbonaceous deposits. Measured amounts of air or oxygen in an inert gas (such as nitrogen) are passed through the catalyst bed, and a careful watch is maintained on the temperature of the combustion front. This should not be allowed to exceed 550°C, so that sintering and loss of molybdenum (which result in an irretrievable loss in surface area and activity) is minimised. The basic reactions involved are the production of carbon dioxide and water from the carbonaceous residues, and of sulphur oxides from the sulphur present in the catalyst.

After regeneration the catalyst is put back on line in the same manner as a fresh catalyst charge.

Future developments

The general principles set out in Chapter 2 (figs. 5 and 6) indicate that developments of future catalysts should be to increase activity, life and strength to optimum values and, if possible, to decrease costs from both raw materials and manufacturing. Future hydro-treating units can be expected to be designed for heavier feedstocks where catalyst regeneration may be frequent. Means of increasing robustness to withstand regeneration, or lowering costs to the extent that fresh catalyst can be used in place of regeneration, are possible developments. Alternatively, catalyst properties which prevent carbon and 'gum' formation, making regeneration unnecessary, clearly deserve close attention.

At high levels of sulphur it is not generally economic to use catalyst 32–4 (zinc oxide) as the sulphur absorber; several alternative systems are therefore in use. These include wash techniques, and, in America, extensive use is made of active carbon. None of the systems has the principle advantage of catalyst 32–4 (related to the stability of ZnS) which is reliability with minimal supervision. An obvious development, therefore, is the use of guard beds of catalyst 32–4 (full use being made of its special pore structure) to follow a variety of alternative absorption system; this also may give scope for the utilisation of the wide temperature range of this catalyst.

The means of obtaining these developments were the subjects of the first three chapters.

CHAPTER 5

Hydrocarbon-reforming catalysts

THE production of hydrogen-containing gases by reacting hydrocarbons with steam in the presence of a catalyst has been an established process since the 1930s. Much of the early development work was done by ICI in conjunction with IG Farben and Standard Oil of New Jersey, and the first ICI reforming plant was commissioned in 1936. It operated at atmospheric pressure with a feedstock of saturated hydrocarbons ranging from methane to butane. During the war, other plants using the same process were erected in North America. These were the forerunners of the many plants now operating throughout the world, most of them at pressures up to 400 lb/in^2 gauge reforming natural gas or the lower saturated hydrocarbons. Many of the early plants used the original catalyst developed by ICI and present catalysts have evolved from this experience.

Before the discovery of natural gas in Holland and under the North Sea, the supply of lower hydrocarbon feedstocks was very limited in Western Europe so, as a result of further research by ICI, the reforming process was extended in 1954 to include hydropetrol, a synthetic petrol made by the high pressure hydrogenation of coal and creosote. The next development was to reform light naphtha, a distillate similar in many ways to hydropetrol, which was becoming available (from the increasing number of petroleum refineries in the world) as a raw material for making hydrogen. The technical problems (notably the removal of sulphur from the feedstock and the development of new catalysts to be able to reform these higher hydrocarbons at pressure without forming carbon) were solved, and in 1959 ICI started up the first naphtha reforming plants. This naphtha process is now used extensively for its original purpose—the production of synthesis gas for ammonia manufacture—but it has also been extended (into the ICI 500 process) to produce town gases up to a calorific value of 500 Btu/ft^3 (4805 kcal/m^3)—a process of considerable value to countries lacking natural gas.

Each stage of the development of the reforming process represented an increase in severity of operating conditions for the catalyst—for methane feedstocks reforming has been at successively higher pressures—for other processes there has been the added complication of higher molecular weight hydrocarbons. One of the most severe limitations in hydrocarbon reforming is the formation of carbon on the catalyst as a result of direct decomposition of the hydrocarbon or of the gaseous products and a theoretical limit is set by the disproportionation of CO. Each stage of the development, in fact, made some improvement in the catalysts necessary. And because gas making is so funda-

The reforming process

When producing gas of comparatively low methane content, the reforming process is endothermic. In the simplest case, with methane feedstock, it follows the course of the reversible reaction

$$CH_4 + H_2O \rightleftharpoons CO + 3H_2 \qquad (1)$$

At the same time, the water gas shift equilibrium is established:

$$CO + H_2O \rightleftharpoons CO_2 + H_2 \qquad (2)$$

A mixture of H_2, CO, CO_2, and CH_4 is consequently obtained, the composition of which is defined mainly by the appropriate chemical equilibria, although it can vary with changes in catalyst activity.

The amount of heat which has to be supplied to the reaction is relatively large so the reforming catalysts (such as catalysts 57–1 or 46–1) are loaded into parallel tubes which are heated in a primary reformer, the exit gas temperature normally being in the range 750–850°C, depending on the gas composition required. To obtain a gas suitable for ammonia synthesis, very low concentrations of methane must be achieved, which necessitate operation at higher temperatures—around 1000°C. Because of metal limitations (especially at pressures which can be over 30 atm) this temperature cannot conveniently be obtained in primary reformer tubes but is practicable in a refractory lined secondary reformer. Here the heat for reforming is produced by the addition of air—which at the same time introduces the nitrogen required for ammonia synthesis. The catalyst in the secondary reformer must be of a refractory type (such as ICI catalyst 54–2). The gas from the secondary reformer (containing H_2, N_2, CO, CO_2 and often less than 0.2 per cent methane) after CO shift, CO_2 removal and methanation, is suitable for ammonia synthesis.

Thermodynamics of reforming

Effect of conditions on product gas composition

The reaction between hydrocarbons and steam is capable of producing a wide range of gases depending upon the conditions of operation of the reformer. Under the conditions normally applicable to reforming, methane is the only hydrocarbon which is thermodynamically stable to any appreciable extent, and so only two reactions (1) and (2) need be considered in determining the equilibrium composition. This is determined by the simultaneous satisfaction of these two equilibria, subject to the constraint imposed by a mass balance. The factors which affect the equilibrium composition are the operating pressure and temperature, the steam ratio (the ratio of moles of steam to atoms of carbon in the feed gas to the reformer) and the carbon to hydrogen ratio of the hydrocarbon feed.

There are two limiting possibilities. Either the whole of the carbon is reformed to give oxides of carbon together with hydrogen or else the reforming proceeds to the limit where no hydrogen is present in the product gas. In the case of reforming a methane feed, this latter limit only occurs when no methane is reformed. But in the case of reforming hydrocarbons higher than methane, it occurs when the maximum amount of carbon is reformed to methane. These two limits correspond respectively to the equilibria attained at very high temperatures, low pressure, and high steam ratios and to the equilibria attained at very low temperatures, high pressures, and low steam ratios. The variation of these parameters is the primary means used to enable steam reforming to be employed for the production of different gas streams. The equilibrium concentrations of CO, CO_2, H_2, and CH_4 as a function of temperature, pressure, and steam ratio for naphtha and methane reforming are given in figs. 29–35. Because the figures cover a very wide range of conditions they are not intended to be absolutely exact; however, the errors should not be greater than ±5 per cent. The method of calculation is described in the appendix.

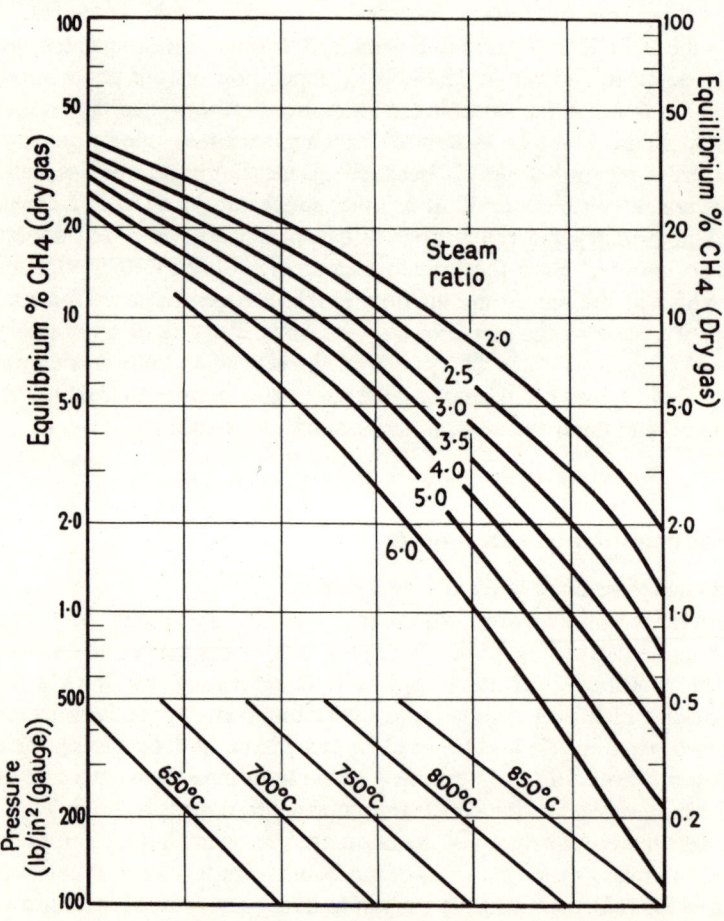

FIG. 29. Equilibrium concentration of methane as a function of temperature, pressure, and steam ratio for methane.

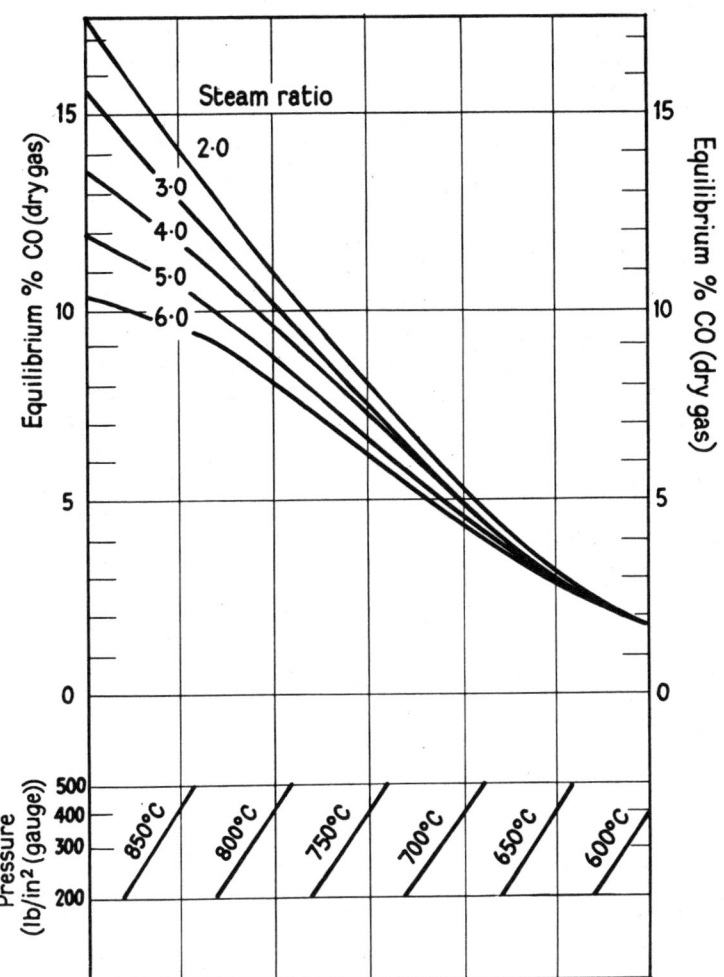

FIG. 30. Equilibrium concentration of carbon monoxide as a function of temperature, pressure, and steam ratio for methane.

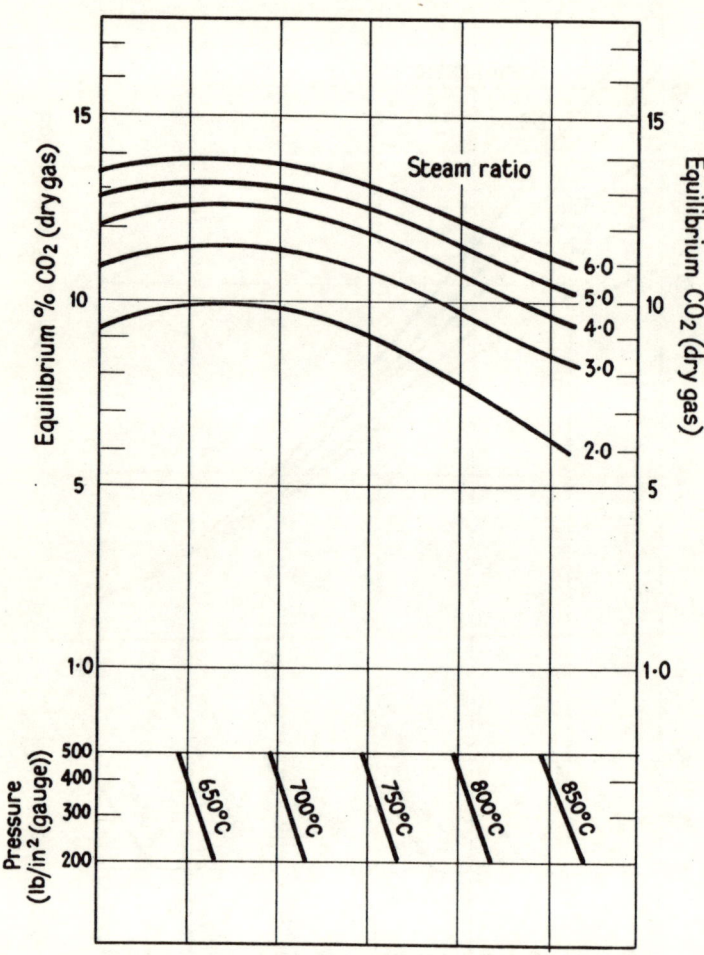

FIG. 31. Equilibrium concentration of carbon dioxide as a function of temperature, pressure, and steam ratio for methane.

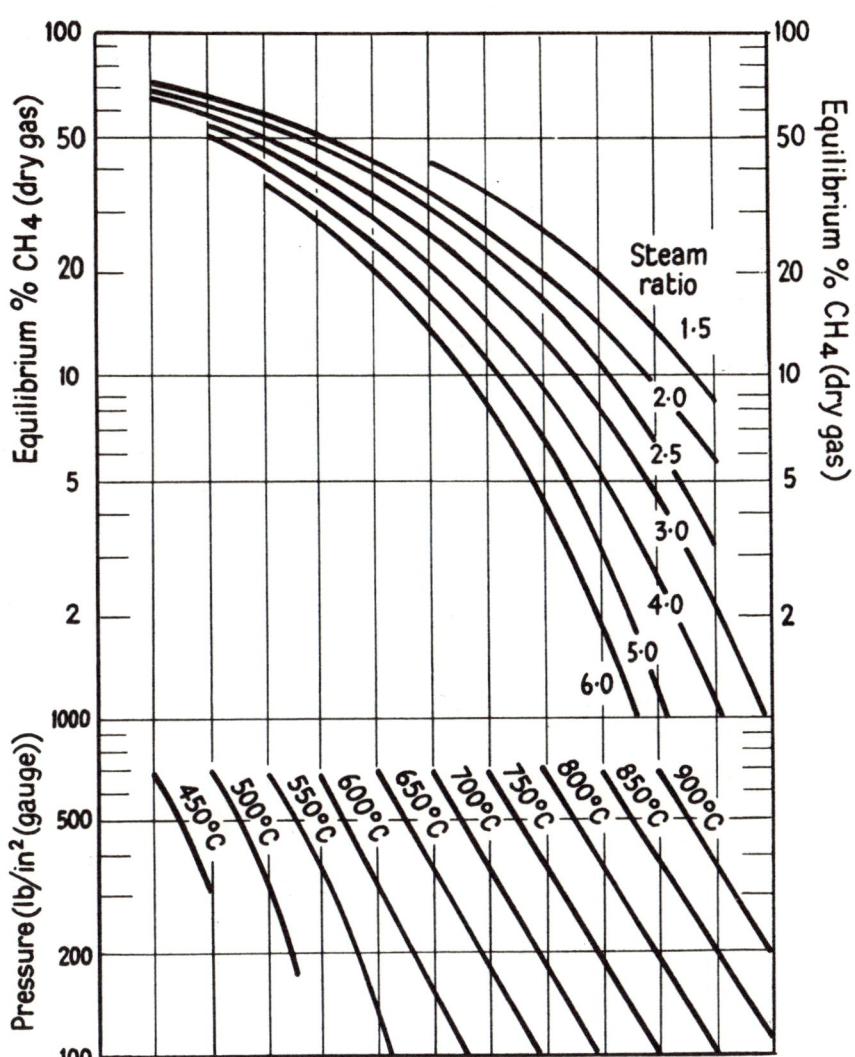

FIG. 32. Equilibrium concentration of methane as a function of temperature, pressure, and steam ratio for naphtha.

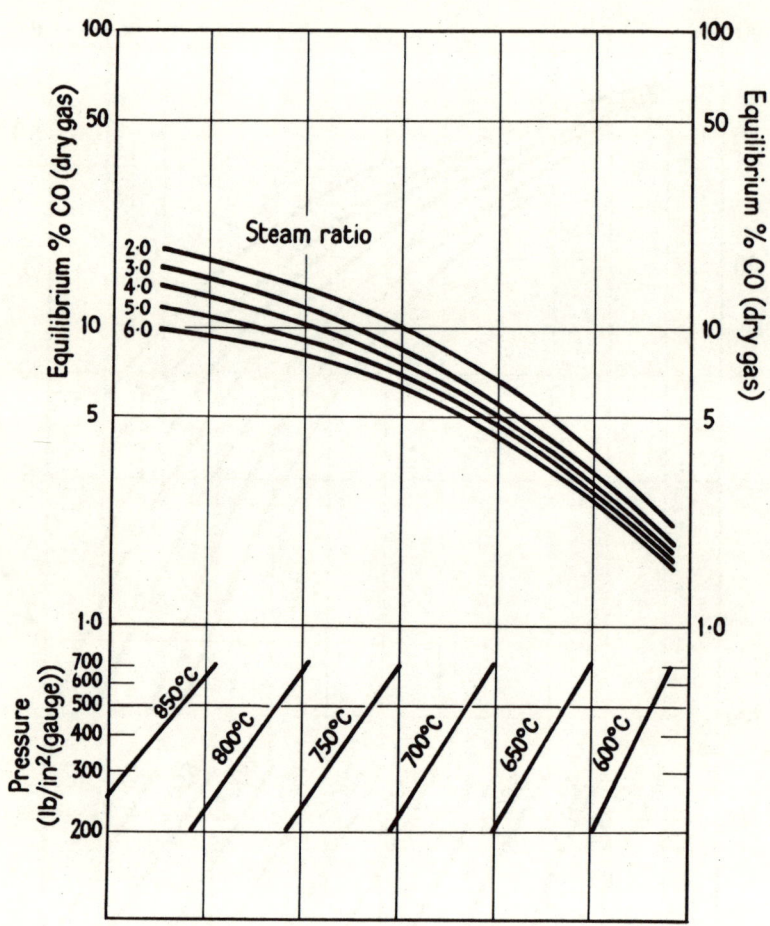

FIG. 33. Equilibrium concentration of carbon monoxide as a function of temperature, pressure, and steam ratio for naphtha.

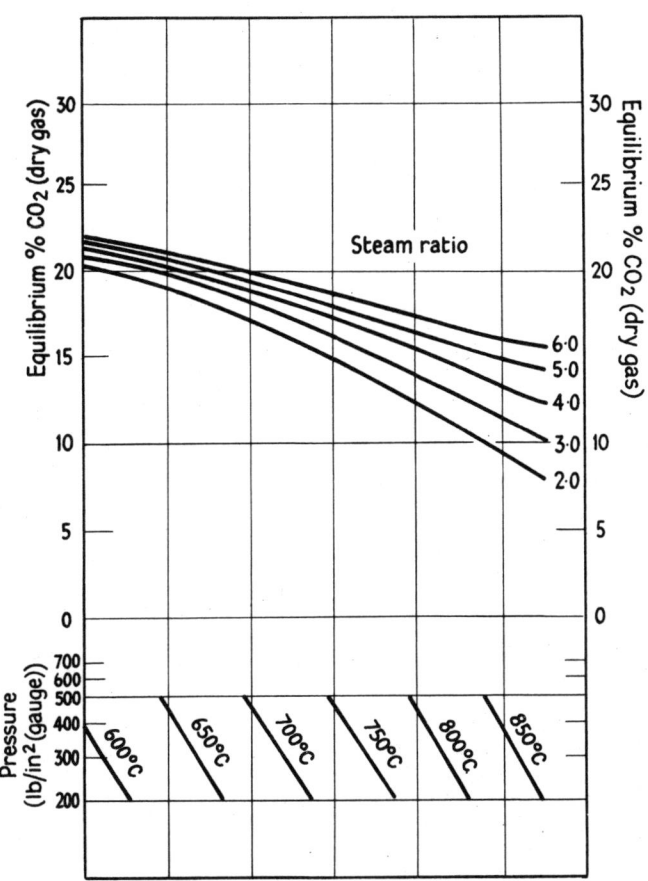

FIG. 34. Equilibrium concentration of carbon dioxide as a function of temperature, pressure, and steam ratio for naphtha.

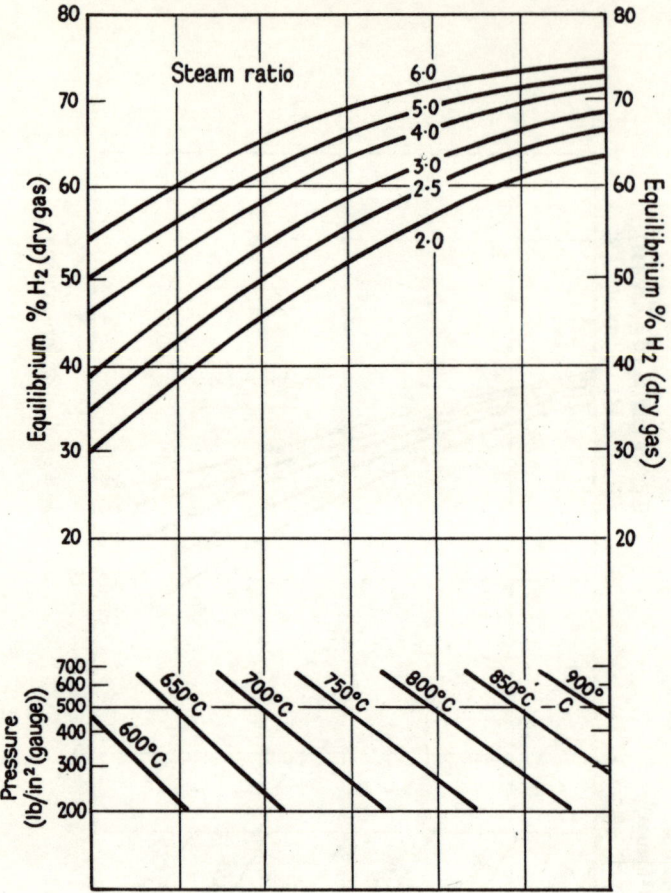

FIG. 35. Equilibrium concentration of hydrogen as a function of temperature, pressure, and steam ratio for naphtha.

Heats of reaction

The enthalpy change for the steam-reforming reaction varies with the reaction conditions.

The methane–steam reaction is always endothermic, as demonstrated by consideration of the reactions

$$CH_4 + H_2O \rightleftharpoons CO + 3H_2 \qquad \Delta H_{25°C} = +49.2 \text{ kcal/mole}$$

$$CH_4 + 2H_2O \rightleftharpoons CO_2 + 4H_2 \qquad \Delta H_{25°C} = +39.4 \text{ kcal/mole}$$

Typical heats of reaction are given below in Fig. 36 for the naphtha–steam reaction proceeding to equilibrium under conditions of practical interest for naphtha $CH_{2.2}$.

The reaction is most endothermic at the limit when the whole of the carbon is reformed to give oxides of carbon together with hydrogen and becomes less endothermic and eventually exothermic as the amount of carbon which is reformed to methane increases.

Conditions			Reaction	$\Delta H_{(25°C)}$
Pressure lb/in^2	Temp. (°C)	Steam ratio		(kcal mole^{-1} CH$_{2\cdot2}$)
300	800	3·0	CH$_{2\cdot2}$+3H$_2$O → 0·2CH$_4$+0·4CO+0·4CO$_2$ +1·94H$_2$+1·81H$_2$O	+24·5
400	750	3·0	CH$_{2\cdot2}$+3H$_2$O → 0·35CH$_4$+0·25CO+0·4CO$_2$ +1·5H$_2$+1·95H$_2$O	+17·9
450	450	2·0	CH$_{2\cdot2}$+2H$_2$O → 0·75CH$_4$+0·25CO$_2$+0·14H$_2$ +1·5H$_2$O	−11·4

FIG. 36. Naphtha reforming heat of reaction.

Formation of carbon

Possible carbon-forming reactions are:

$$2CO \rightleftharpoons C + CO_2 \qquad (3)$$

$$CO + H_2 \rightleftharpoons C + H_2O \qquad (4)$$

$$CH_4 \rightleftharpoons C + 2H_2 \qquad (5)$$

and in naphtha reforming

$$\text{pyrolysis of higher hydrocarbons} \qquad (6)$$

Calculations can be carried out to define a thermodynamic minimum steam ratio if the equilibrium constants for the carbon forming reactions are known. Dent has published (ref. 29) thermodynamic data for carbon forming reactions (3), (4), and (5) on nickel. The method of calculation involves a determination of the minimum steam ratio below which the presence of carbon is inevitable if the gas is brought to thermodynamic equilibrium. The results of these calculations based on the Dent data for naphtha reforming and methane reforming over a range of pressures and temperatures are presented in figs. 37a and 37b.

Whilst these calculations establish the minimum steam ratio below which carbon will be formed if the system is brought to equilibrium, they give no information about the possibility of carbon formation if the system is not at equilibrium. The gas composition in a reformer can be such that thermodynamic theory predicts the formation of carbon by one of the carbon-forming reactions —(6) for example—and its removal by another. The question of whether there is a net build-up of deposited carbon is then a kinetic one. The availability of suitable kinetic data is limited, and results obtained by different people vary widely due to differences in the form of the carbon used and in the surfaces upon which carbon was deposited. Some general guides to the rates of the various reactions can be obtained from the literature.

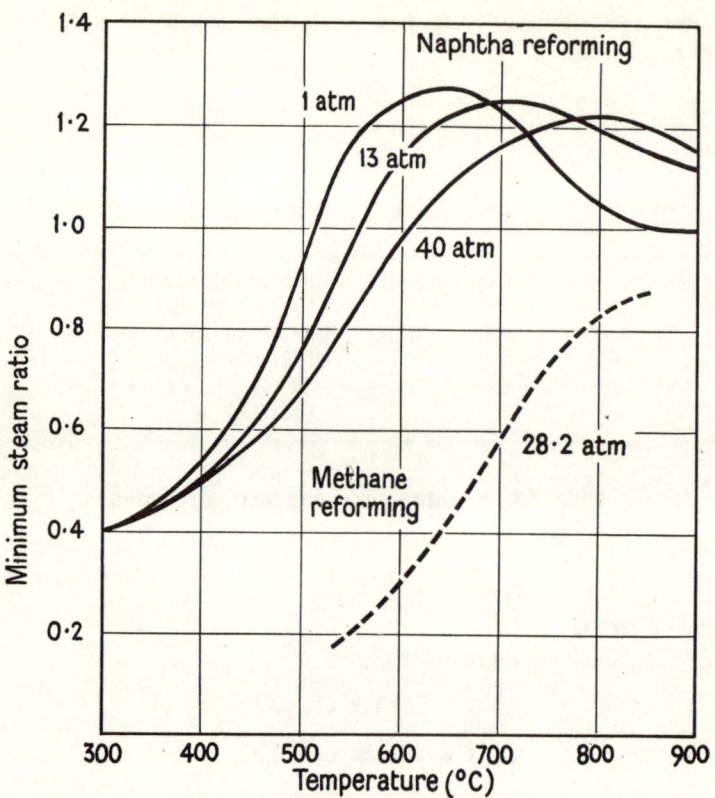

FIG. 37. (a) Thermodynamic minimum steam ratios. (b) Equilibrium constants for carbon-forming reactions.

(1) $CO + H_2 \rightleftharpoons C + H_2O \qquad K_p = \dfrac{P_{H_2O}}{P_{CO} P_{H_2}}$

(2) $2CO \rightleftharpoons C + CO_2 \qquad K_p = \dfrac{P_{CO_2}}{P_{CO}^2}$

(3) $CH_4 \rightleftharpoons C + 2H_2 \qquad K_p = \dfrac{P_{H_2}^2}{P_{CH_4}}$

Rates of carbon formation

Carbon formation from CO is three to ten times faster than from CH_4 under the same conditions (ref. 30).

Butane deposits carbon at about the same rate as CO (ref. 30).

Benzene and toluene deposit carbon at a rate several orders of magnitude faster than CH_4 (ref. 31).

Rates of carbon removal

Carbon removal by CO_2 is about ten times faster than its formation from CO (ref. 32).

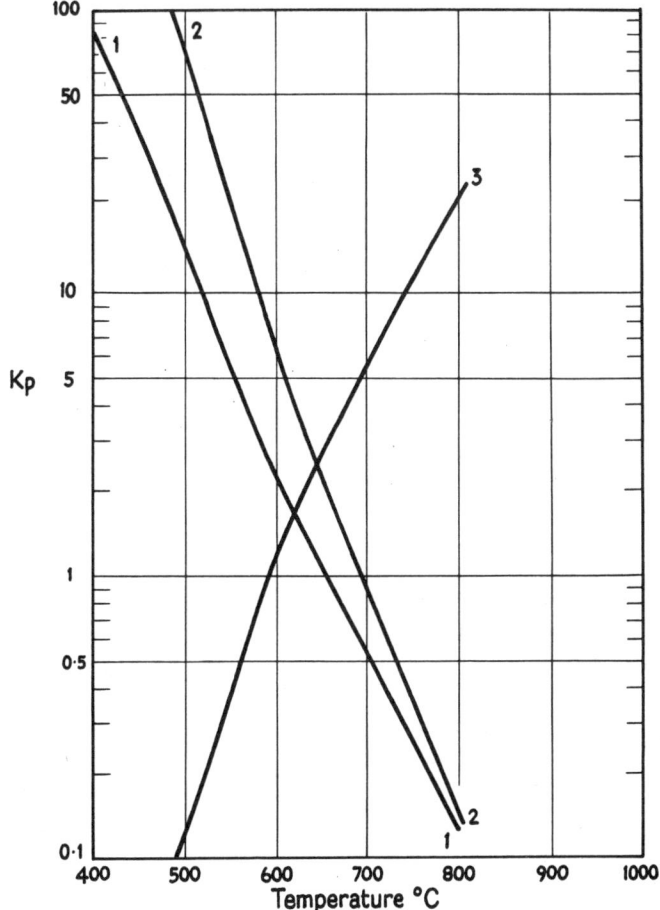

FIG. 37. (b) (Explanation on facing page.)

The C–H_2O reaction is two to three times faster than the C–CO_2 reaction, but the C–H_2 reaction is 1000 times slower (refs. 33, 34, 35).

The rates of the C–H_2O and C–CO_2 reactions are increased by a factor of 100 or more by small amounts of alkali metal salts of weak acids (refs. 36, 37).

In methane reforming only three reactions need be considered: (3), (4), and (5). Reaction (4) is obtained by subtracting the water gas shift reaction (2) from reaction (3). Since the water gas shift reaction is always virtually at equilibrium in a reformer, so reaction (4) will always be on the same side of equilibrium with respect to carbon as reaction (3). From the given kinetic data, the rate of establishment of equilibria (3) and (4) is faster than that of equilibrium (5). Hence the thermodynamics of reaction (3) are most important and if P_{CO_2}/P_{CO}^2 is less than the equilibrium value at any point in the tube then carbon deposition is inevitable in time, irrespective of the position of other equilibria. The steam ratio at which this occurs is the thermodynamic minimum steam ratio, previously calculated.

In the case of naphtha reforming, reaction (6) must also be considered. The pyrolysis reactions are at least as fast as reactions (3), (4), and (5). The heavier the feedstock and the more unsaturated compounds it contains the higher will be the practical minimum steam ratio. Doping the catalyst with alkali metal salts allows a closer approach to the thermodynamic minimum steam ratio calculated from reactions (3), (4), and (5), as it greatly enhances the rates of the $C-H_2O$ and $C-CO_2$ reactions. There is a point in the first few feet of a primary reformer at which the driving force for carbon removal is at its lowest value, and this is where carbon is most likely to be deposited, particularly since the concentration of higher hydrocarbons is still appreciable in this region.

Secondary reforming

For ammonia synthesis gas production the necessary nitrogen content is introduced as air into a secondary reformer, a catalyst-filled vessel in which the oxygen of the air is burnt, liberating heat and raising the temperature of the product gases from the primary reformer. A secondary reformer product–gas equilibrium, at a temperature some 150° to 200°C higher than that of the

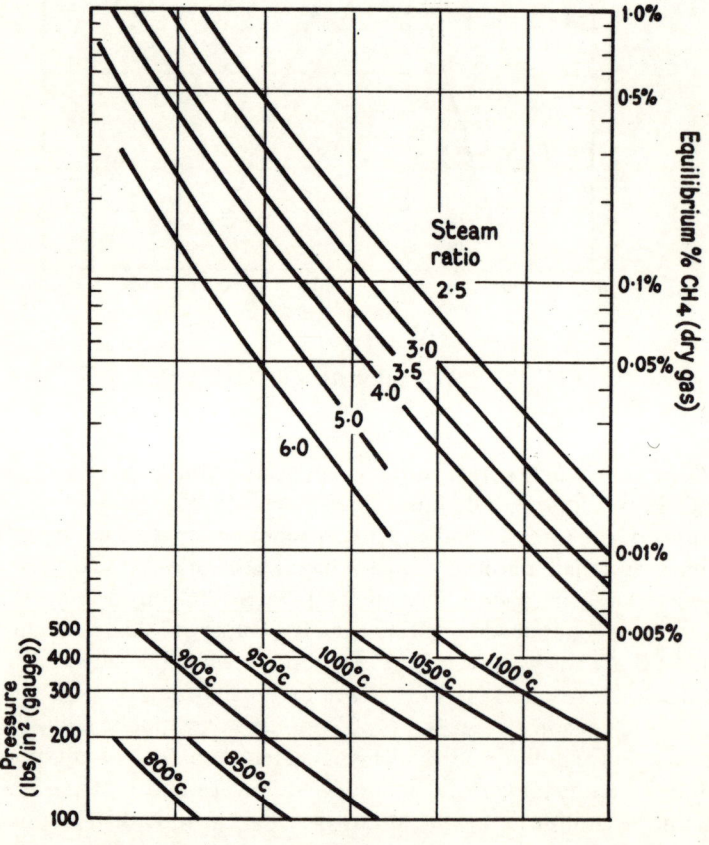

FIG. 38. Equilibrium data for methane in secondary reformer of ammonia plant. Feed, $CH_{2.3}$; O_2/C molar ratio, 0.22.

product gases from the primary reformer, is thus achieved. At this higher temperature the equilibrium methane-content of the gas is decreased.

The final equilibrium is influenced by the operating pressure, steam ratio, the air-to-carbon ratio, and the final temperature. The air-to-carbon ratio used will be determined by the requirement that the final gas should contain H_2 and N_2 in a 3:1 molar ratio, and since the reactor is operated adiabatically the final temperature will then be fixed by the inlet conditions. Some typical equilibrium data are provided in fig. 38 for a range of pressures, temperatures, steam ratios, and a given air/carbon ratio.

Effect of CO_2 on equilibrium gas-composition

The production of high-CO-containing gases can be achieved by the recycle of CO_2 to the reformer. This procedure is adopted when running a reformer to make methanol synthesis gas ($CO:H_2$, 1:2) and 'oxo' synthesis gas ($CO:H_2$, 1:1).

The effect of additional CO_2 on the primary reformer equilibrium-gas-composition is illustrated for a few cases in fig. 39.

Reformer feed				Reformer conditions		Equilibrium gas composition (% v/v (dry gas))			
Naphtha $CH_{2.3}$ (moles)	Methane CH_4 (moles)	Steam H_2O (moles)	CO_2 (moles)	Pressure lb/in^2	Exit temp. (°C)	CH_4	CO	CO_2	H_2
1	—	3	3	400	800	1·8	24·8	48·2	25·2
1	—	3	3	400	850	0·7	28·0	45·4	25·9
1	—	3	3	200	800	0·7	26·2	46·1	27·0
1	—	3	3	200	850	0·2	28·6	44·5	26·7
1	—	2·5	2·5	400	800	2·7	26·6	44·0	26·7
1	—	2·5	2·5	400	850	1·2	30·3	40·5	28·0
1	—	2	2	400	800	4·2	28·6	39·2	28·0
1	—	2	2	400	850	2·0	32·8	34·9	30·3
1	—	2	2	200	800	1·9	31·0	35·7	31·4
1	—	2	2	200	850	0·7	34·1	32·9	32·3
—	1	2	2	200	800	3·3	29·5	28·3	38·9
—	1	2	2	200	850	1·4	32·8	25·1	40·7
—	1	3	3	400	800	3·3	25·0	40·7	31·0
—	1	3	3	400	850	1·4	28·7	37·1	32·8
—	1	5	5	400	850	0·4	23·6	51·1	25·0

FIG. 39. CO_2 reforming equilibrium—gas composition.

Reforming catalysts

Different catalysts are required for the different duties of primary and secondary reforming, and for different feedstocks. Nickel has been found to be the most effective metal for the reforming of hydrocarbons, and it is the active component in most of the catalyst formulations which are available. These formulations differ mainly in respect of the other components that are present, such as alumina, magnesia, calcium oxide, and so on. These components have an influence on the catalytic properties as well as on such physical properties as strength, density, or refractoriness.

The three main catalysts which ICI supplies for steam reforming processes are described in fig. 40 and typical analyses are given in fig. 41.

Catalyst	Duty
57–1	Primary reforming of natural gas and methane at pressures up to at least 30 atm and temperatures up to 850°C (and for gaseous saturated hydrocarbons)
46–1	Primary reforming of hydrocarbons higher than methane at pressures up to 30 atm and temperatures up to 850°C.
54–2	Secondary reforming of primary gases at pressures up to at least 30 atm and temperatures up to 1300°C.

FIG. 40. Reforming catalysts.

Component %	57–1	46–1	54–2
NiO	32	21	18
CaO	14	11	15
SiO_2	0·1	16	0·1
Al_2O_3	54	32	67
MgO	—	13	—
K_2O	—	7	—

FIG. 41. Reforming catalysts, typical analyses (loss free basis).

The function of the various components in these catalysts is discussed in the following sections.

Nickel

Nickel metal is the catalytic species in all commercial catalysts. Cobalt can be used but it is less effective and it is also more expensive. The precious metals, Pt, Pd, Ir, Rh, Ru, and so on, are more active per unit weight than nickel (refs. 38, 39). But their cost per unit of activity must be considered, and their advantages then disappear on economic grounds. They can be used in conjunction with nickel (refs. 40, 41, 42). The use of cobalt or of precious metals does not completely overcome the difficulties due to carbon formation from the higher hydrocarbons, as equilibrium considerations rather than catalytic properties dictate the temperature of the reforming process, and thermocracking of the feedstock must still occur.

Thus nickel is the recognised element. It is normally present in the manufactured form of the catalyst as nickel oxide which is reduced to the metal in a reformer immediately before use. The application of nickel to the reforming process, in combination with various support materials (refs. 43, 44, 45, 46), has been the object of much research. As with most catalysts, the maintenance of catalytic effectiveness over long periods is the main requirement. High activity is associated with a high surface area of nickel. This is obtained with a small crystal size, and growth of the crystals must be prevented, or at least retarded as much as possible. In steam reforming at temperatures above 750°C, and in the presence of high partial pressures of steam, conditions encourage crystal growth (Chapter 2). Preservation of the nickel surface area, therefore, is one of the main functions of the support materials which, as described in pages 23–24, are selected for their refractoriness, and similar properties. That is to say, they act as 'stabilisers'.

The stabilisers and support materials in the ICI catalysts have been selected to provide the best physical properties commensurate with the requirements of minimising the loss of nickel surface area in use. The surface area of the nickel metal in a reforming catalyst in use is normally about $0.5\ m^2/g$ of nickel, depending on the operating conditions, which corresponds to a crystallite size of about 1 micron.

Catalyst supports

The operating conditions for reforming catalysts are very severe. The temperatures, even for primary reforming, are some 200–300°C higher than the temperatures at which most other catalysts operate and in addition the catalysts are exposed to high partial pressures of steam and hydrogen. Physical properties are very important and the formulation must be such as to give a product which possesses maximum strength throughout its life. At the same time, the catalytic properties (such as activity and selectivity) must not be impaired by these physical limitations.

Excellent physical properties can be obtained by using a simple support material, such as alumina or magnesia, that has been fired at a temperature around 1500°C. The support can be in the form of lumps, or pre-formed pellets, or rings, which are impregnated with nickel by soaking them in a nickel salt solution and then drying them—an operation which can be repeated to obtain the required nickel content. The influence of the support material on catalytic behaviour was discussed in Chapter 2 (pp. 24 and 25) where it was shown that the crystal size of the support material largely determines the extent to which activity can be stabilised. These impregnated catalysts usually exhibit comparatively low activity, particularly if activity in service rather than initial activity is measured. Some reforming catalysts are made in this way for certain limited applications. Being based on fired alumina or magnesia, they are very refractory and can therefore be used as a layer on top of a more conventional catalyst in a secondary reformer. This is the point where, as a result of air injection, the highest catalyst temperatures are attained. Given a good refractory secondary catalyst (such as 54–2), however, the low-activity type of impregnated catalyst is not necessary.

In the more conventional type of primary-reforming catalyst, physical strength is obtained by incorporating a hydraulic cement among the ingredients. As indicated later, the silica content of these catalysts usually has to be very low and therefore, instead of a Portland cement, a calcium aluminate type is normally used. The cement is used to bond together the other components—which in the simplest case would be nickel oxide and alumina.

Magnesia is also included in some formulations and, being more basic than alumina as well as refractory, it can be advantageous. It has to be used with caution because it can hydrate under some conditions in a reformer, particularly at start-up and shut-down, leading to weakening and disintegration of the catalyst. The calculated values for the decomposition partial pressures of water for the reaction $Mg(OH)_2 = MgO + H_2O$ at different temperatures are given in fig. 42. The value is 30 atm at about 425°C so that, in a reformer at this pressure, hydration would occur at temperatures below 425°C during a start-up with circulating steam. In pressure reformers using catalysts containing free magnesia, contact with steam below about 500°C must be avoided. Similar considerations apply to calcium oxide, although in most catalysts the CaO is combined, for example, as calcium aluminate in the cement. This is the case with the ICI reforming catalysts—all contain CaO and the naphtha reforming catalyst also contains MgO, but both are combined with other components and consequently do not hydrate and the stability of the catalysts is thus preserved.

Temperature (°C)	250	300	350	380	400	450	500	600
Partial pressure of water, atm	0·32	2	8	14	22	50	140	630

FIG. 42. Decomposition partial pressure of $Mg(OH)_2$.

The earliest catalyst for reforming hydrocarbons at atmospheric pressure contained acidic components, notably silica. This was introduced into the catalyst as kaolin or china clay, that is to say as aluminium silicates, which in the finely divided state act as binders, and also facilitate the compaction and cohesion of the solid during manufacture.

At the higher pressure of modern reforming processes, the volatility of silica in steam is significant, so that it is slowly removed from these catalysts and is deposited in cooler parts of the plant such as boilers and heat exchangers downstream of the reformer. Data on the solubility of quartz and vitreous silica in steam at temperatures up to 600°C have been reported (refs. 47, 48). In work with reforming catalysts at ICI, experiments have been undertaken with various siliceous materials at temperatures up to 850°C and at pressures up to 400 lb/in^2. The results with vitreous silica (which was used in the earlier catalysts, and which is present in many refractory materials used for reformer linings) are given in figs. 43 and 44. These extend the data cited above (refs. 47, 48) and show that the concentration of silica in steam is proportional to steam pressure. This is consistent with silica being volatilised as ortho silicic acid, $Si(OH)_4$. The results do not support the view sometimes expressed that the transfer occurs as SiO produced by the reduction of SiO_2.

The concentration of silica in steam which has passed over vitreous silica at

Temperature (°C)	300	400	500	600	700	800	900
SiO_2 concentration (ppm w/w)	0·24	1·00	3·2	7·0	15·0	26·0	40·0

FIG. 43. Effect of temperature on transfer of silica in steam, at 180 lb/in^2 gauge (vitreous silica).

Pressure (lb/in^2 gauge)	30	50	100	200	300	400	500
SiO_2 concentration (ppm w/w)	3·3	5·2	10·7	21·4	31·6	41·7	52·5

FIG. 44. Effect of pressure on transfer of silica in steam, at 750°C (vitreous silica).

750°C and 200 lb/in^2, is about 20 ppm by weight. The quantity volatilised from a typical fire brick (51 per cent SiO_2) is about 5 ppm but a natural gas reforming catalyst containing below 0·3 per cent SiO_2 gives rise to a concentration less than 0·03 ppm. Silica evolution is not a problem encountered with any of the catalysts 57–1, 54–2 or 46–1. Catalyst 46–1 contains silica, but it is firmly combined with the alkali, and the other basic materials MgO and CaO that are incorporated. In this combined state it plays an important part in the functioning of the catalyst but is not volatile. Materials used for the refractory linings for vessels and pipes and so on still offer potential sources of silica so particular care is needed during their selection.

Methane-reforming catalysts

The formulation of a reforming catalyst is very much influenced by the operating conditions which in turn are affected to some extent by the feedstock to be reformed. Because of the increased tendency to form carbon, the reforming of naphtha requires a special formulation, as discussed later on (pp. 82–83). Carbon formation is not a problem with methane reforming catalysts but there are many other factors to be taken into account in devising the ideal formulation.

The general structural properties of catalysts have been discussed in Chapter 2 which stressed the requirements of high activity, long catalyst life, high physical strength and shape (affecting gas flow). Another factor is the production cost of the final catalyst which is controlled by choice of raw materials and method of manufacture. Some of these requirements conflict with each other, e.g. activity and physical properties. Thus in a reforming catalyst an increase in the nickel content raises the activity but this can adversely affect the physical properties as well as increasing the cost. High activity is related to the surface area of the nickel, and long catalyst life (in terms of activity) is achieved by preserving the high surface area. This can be done during manufacture when selected support materials can be suitably compounded to provide a stable, non-sintering surface for the active material. At the same time, the structure of the support donates the mechanical properties to the catalyst which must prevent physical disintegration. Clearly the support must undergo no phase changes nor be affected by any of the reforming reactants or products, notably steam. For example, as indicated on p. 80, silica cannot be used although it would otherwise contribute excellent physical strength to the catalyst. Similarly, support materials prepared at high temperature, although strong physically and stable, do not have the adequate surface areas which preserve a high area of nickel surface. The creation of a high area support which can also retain high physical strength during a long lifetime therefore presents special problems to the manufacturer.

The conflicting requirements of activity, strength, life and so on can be satisfied (as in catalyst 57–1) by a catalyst formulation which utilises the extensive knowledge of the raw materials and of the solid-solid and solid-gas reactions occurring in the catalyst in a reformer. Full value is obtained, in catalyst 57–1, from the nickel content, which is high compared with a naphtha reforming catalyst, by using a special alumina as the stabilising medium. And another feature of the formulation is the inclusion of some calcium oxide. The formation of calcium aluminates in the catalyst support, during use, maintains the high physical strength of catalyst 57–1, and this is achieved without any loss of activity.

Naphtha-reforming catalysts: alkali and catalyst acidity

The formulation of a catalyst to reform naphtha is further complicated by the greater tendency of the higher hydrocarbons to form carbon. The thermodynamics of the carbon-forming reactions have been discussed on pages 73–74. The carbon can be formed in several ways—by cracking of the hydrocarbons, either homogeneously, or catalytically on the active nickel surface or on the catalyst support material. The catalytic effect of acidic oxides, such as aluminosilicates, on the cracking of hydrocarbons, is well known in the petroleum industry. The effect is also obtained with amphoteric oxides, and with some compounds normally regarded as basic but which, at high temperature and high partial pressures of steam, take on some acidic characteristics.

The cracking activity presents no difficulties in the reforming of methane, and hydrocarbons below butane, but catalysts have to be modified to obtain satisfactory operation with naphtha at pressures above atmospheric. Catalyst 46–1 is one in which the problems of activity and carbon formation have been dealt with successfully. The use of basic support materials (such as magnesia or magnesia-alumina spinels) to support the nickel has been tried but is not effective, partly because a truly basic support is difficult to achieve in the presence of high pressure steam, and partly because nickel itself gives rise to hydrocarbon cracking, and hence to carbon formation. Such a catalyst would operate under test conditions at a steam/carbon ratio of 3·5–4·0, compared with 10·0 for an 'acidic' catalyst of the early type. But this does not permit economic plant running, because reliable operation (in practical terms) at these ratios requires the catalyst to operate at a ratio of around 2·5–3·0 under test conditions. To obtain an adequate margin of safety, for continuous plant operation at 3·0 steam ratio, a catalyst which will function without forming carbon (under test conditions) at a steam ratio of 1·5 is required.

Catalyst 46–1 will operate for long periods at these low steam ratios in full-scale plants. In semi-technical experiments it has operated near the thermodynamic minimum ratio of 1·2 for several weeks (fig. 37). As described in numerous patents (refs. 49–53), this performance is achieved by the presence of an alkali compound, such as the hydroxide or carbonate of sodium or potassium, which play a complex role in the functioning of the catalyst. The potash in catalyst 46–1 neutralises acidity in the catalyst support, reacting with aluminosilicates which are present in the clays used in its manufacture, to form complex potassium compounds such as kalsilite, $KAlSiO_4$. At reforming temperatures these complexes, under the action of high-pressure steam and CO_2, decompose slowly, releasing low concentrations of potash, which is an effective catalyst for the reaction of steam with any incipient carbon that may have been produced by non-acidic routes—that is by decomposition of the hydrocarbons, either thermally, or catalytically by contact with nickel metal.

The formation of the potassium complexes in the catalyst during manufacture is essential, as it is their slow decomposition, while the catalyst is in use in a reformer, which produces free alkali at the required rate to prevent carbon formation. The released potash is mobile, and hence is effectively distributed throughout the catalyst. For the same reason, it is slowly lost from the catalyst during use, so that sufficient must be initially present in the combined state to ensure adequate life. Catalyst 46–1 contains about 7 per cent K_2O. In many plants it has been in use for longer than three years.

The alumina and silica, which are produced by the decomposition of the potassium complexes, react with magnesia and lime forming spinels and compounds such as montecellite—$CaO.MgO.SiO_2$. Magnesia is added to the catalyst for this purpose, as well as for its general refractoriness. At all times, even in unused catalyst, the magnesia is combined, so that hydrolysis by high pressure steam is not a problem.

Other catalysts in which nickel is supported on basic materials, such as magnesia alone or magnesia alumina spinels, have been proposed for naphtha reforming at pressure, because, when treated with alkali, such catalysts will operate at low steam/carbon ratios in a manner comparable to 46–1. They will, however, only do so for relatively short periods for after one or two months (instead of one to three years or even longer) carbon is produced. The complexes which provide the slow release of alkali from 46–1 are not formed in these other catalysts. Instead, alkali is evolved for a short time, at a higher rate than is necessary to prevent carbon formation, with the result that the catalyst quickly becomes depleted and carbon is deposited.

Catalyst poisons

Like most catalysts, nickel reforming catalysts are very sensitive to even the low concentrations of certain impurities that can be present in the feedstocks. The elements most often encountered are sulphur, arsenic, halogens, phosphorus and lead. Some cause permanent damage to the catalyst. Others have only a temporary effect, catalytic activity returning to normal when feedstock purity is restored.

Sulphur

Sulphur is often present in natural gas, generally as H_2S (concentrations are usually low and rarely exceed 300 ppm v/v). The light naphthas used as reformer feedstocks contain sulphur in concentrations up to 500 ppm w/w, occasionally up to 1500 ppm w/w. Much of the sulphur in naphtha is in the form of organic compounds such as sulphides, disulphides, mercaptans and ring compounds. These can all be removed by the desulphurisation processes described in Chapter 4. If (as a result of poor operation) H_2S or any of these compounds passes through the desulphuriser, the nickel reforming catalyst is rapidly poisoned, the extent depending upon the amount of H_2S present. All sulphur compounds are readily hydrogenated to H_2S over a nickel catalyst at reforming temperatures.

Small amounts of sulphur can have a significant effect on catalyst activity although the effect varies from catalyst to catalyst. The loss in activity due to small amounts of sulphur appears to be greater on modern, more active, catalysts than has been reported by earlier workers. In a reformer operating at a catalyst exit temperature of 750°C, the catalysts 57–1 and 46–1 can only show their full activity (which is high) when the sulphur concentration in the feedstock (methane or naphtha) is below 0·5 ppm S w/w. Much work has been done to establish the concentration of sulphur that can be tolerated by nickel catalysts (refs. 54, 55). These researches have thrown light on the poisoning reaction and have indicated that the tolerable level of sulphur in the feedstock is closely related to the catalyst's activity as well as operating conditions and so on.

ICI's recent work on the poisoning of the naphtha reforming catalyst 46–1 illustrates the effect of sulphur. Sulphur lowers the activity of the catalyst, the level of the activity decreasing with increasing sulphur content of the feedstock. The rate of die-off increases when sulphur concentration is raised. The poisoning effect of sulphur is reversible and at any given operating temperature there is a sulphur concentration below which no detectable poisoning occurs. A poisoned catalyst soon recovers its full activity when operated with feedstock containing sulphur in concentrations below that level. The sensitivity of the catalyst to poisoning is increased when operating temperatures are lowered. Thus, in small-scale isothermal experiments reforming heptane at 750°C on 46–1, the tolerable sulphur level was found to be about 0·6 ppm S w/w. At 700°C it is about 0·2 ppm.

Similar experiments have been carried out with methane-reforming catalysts. Morita and Inoue (ref. 54) give data on the effect of temperature on the threshold concentration of sulphur compounds (mercaptan or thiophene) in experiments at 800°C and above with catalysts containing 5 to 20 per cent nickel. Pichler surveys data from many sources relating to many different catalysts under different operating conditions (ref. 55). The figures are summarised in fig. 45.

There is a wide disparity between the results, which is no doubt due to the difference in catalysts and test conditions. The two catalysts examined by Morita and Inoue show the effect of nickel content on the sulphur tolerance. The high figures given by Pichler indicate that the catalysts in his survey were relatively inactive. The lowest values are the ICI figures, and these are relevant to full-scale reformers operating at the indicated exit temperatures. Catalyst

Catalyst	Catalyst exit temperature, T(°C)			Remarks
	800	850	900	
ICI (14–25 per cent Ni)	1	5	25	Full-scale reforming; catalyst temperature $400°-T°C$
Morita and Inoue (25 per cent Ni	4	16	—	Isothermal beds
5 per cent Ni)	—	2	—	
Pichler (various catalysts)	60	76	120	Not defined

FIG. 45. Tolerable sulphur concentrations in methane (mgm S/Nm³).

which is nearer the reformer inlet is at a lower temperature, and poisoning is consequently noted at lower sulphur concentrations than those established by the isothermal experiments.

Sulphur is more often encountered in naphtha than in methane and considerable experience in the use of catalyst 46–1 with full-scale reformers (in addition to semi-technical experiments) has made it possible to define sulphur concentrations in light naphtha which will, in technical practice, ensure operation without loss of activity. At 750°C catalyst exit temperature the maximum tolerable concentration is 0.5 ppm S in naphtha w/w. The ICI desulphurisation process is designed to give sulphur concentrations well below this. The tolerable concentration is higher at higher temperatures, but it is desirable to keep the sulphur concentration below 0.5 ppm at all times.

Poisoning of a nickel catalyst must be associated with reaction between sulphur and the active nickel surface. Only small concentrations of sulphur are required to poison the catalyst so that the formation of bulk sulphide by the reaction $3Ni + 2H_2S = Ni_3S_2 + 2H_2$ is not involved. Over the temperature range of a typical reformer (400–800°C) bulk sulphide formation by that reaction requires the value of P_{H_2S}/P_{H_2} to exceed 1.0–10.0 (ref. 56). In practice the hydrogen concentration in a reformer is high—particularly at the exit—and yet poisoning occurs at values of P_{H_2S}/P_{H_2} of 0.01 or less. The amounts of nickel and of sulphur which react are very small. A reforming catalyst containing 15 per cent Ni operating at an exit temperature of 775°C is poisoned when it contains only 0.005 per cent S, corresponding to the sulphiding of only 0.06 per cent of the nickel. This is equivalent to poisoning all the nickel atoms in the surface of crystallites 1 micron in diameter. At lower temperatures (around 600°C) the values are a factor of ten lower.

From these observations it is clear that at the top of a reformer, where the temperature is around 400–500°C, the active parts of the catalyst can be sulphided even with a feedstock desulphurised to the specified low level. This is of no practical significance, however, because the amount of catalyst involved is very small (perhaps 1–2 per cent of the catalyst in the reformer) and also because the temperature is rising rapidly in this region. If, however, the desulphuriser performance deteriorates, due for example to maloperation, the sulphided zone will extent further down the catalyst bed until the performance of the reformer will be impaired.

The falling off in performance would be apparent as an increase in methane slip, and, from a naphtha reformer, as an increase in the aromatic compounds in the product gas. This can be rectified by removing the sulphur contaminant from the feedstock, after which the catalyst performance will return to normal in a few days. The recovery can be hastened by steaming the catalyst without feedstock for 12–24 hours. When reforming the higher hydrocarbons, poisoning should be rectified as soon as possible, because poisoning can upset the kinetic balance between the carbon-forming and carbon-removing reactions and carbon can be deposited on the catalyst. This aggravates the decrease in activity leading to an increased pressure drop, and, eventually, to high reformer-tube temperatures. The carbon can be removed by steaming without harming the catalyst but its formation is to be avoided if possible.

Most catalysts contain some sulphur (up to 0·03 per cent S) which is unavoidably introduced during manufacture, usually in the raw materials. An important function of the starting-up procedure for a reformer, during which the catalyst is heated under steam and hydrogen, is the removal of this sulphur. Using 1–2 kg steam/kg of catalyst and about 0·25 $m^3 H_2$, the ICI catalysts can be desulphurised in one day or less because of their low sulphur content. The desulphurisation usually takes place during the normal warming up and start-up procedure. The desulphurisation need not be completed before feedstock is introduced because sulphur removal continues during normal operations. High throughputs cannot be employed until all sulphur has been removed and the catalyst has reached its maximum activity.

Arsenic

Another catalyst poison of practical significance is arsenic which, on some ammonia synthesis plants, can originate from the CO_2 removal process where arsenic is used as a catalyst. Contamination of the steam passing to a primary reformer can sometimes occur as a result of mechanical failures in the CO_2 plant. The effect is the same as with sulphur, loss of performance being apparent on a natural gas catalyst as an increase in methane slip, and by the appearance of hot tubes, particularly at the top (inlet). On naphtha reformers, the aromatics content and the ethane content of the product gas increase. Samples of poisoned catalyst from the top of full-scale reformer tubes have been found to contain up to 1000 ppm As_2O_3. Samples from the bottom half of the tube are often uncontaminated, containing below 10 ppm As_2O_3 (as in new catalyst). Experiments on arsenic poisoning have been carried out by ICI in semi-technical naphtha reformers. These showed that the effect of arsenic is apparent when the As_2O_3 content of the catalyst (46–1) exceeds 50 ppm.

For practical purposes, unlike sulphur, arsenic poisoning is not reversible, and therefore the threshold concentration of arsenic in the reactants below which poisoning does not occur is very low. Arsenic present in any concentration will accumulate on the catalyst until it produces a detectable effect. As with sulphur, the effect will not be apparent until poisoning has affected the small proportion of catalyst which is at the low temperatures near the inlet of the reformer. The action is delayed while the poison accumulates in the catalyst and on the plant itself. In one experiment where naphtha was reformed on catalyst 46–1, at 20 atm pressure with a catalyst exit-temperature of 785°C, the effect of 1 ppm As_2O_3 in the steam was not apparent until after four days. In that experiment, some of the arsenic impregnated the reformer tube, and was transferred to fresh catalyst later charged to the same tube, causing that catalyst also to be poisoned. In the event of serious arsenic poisoning, the reformer tubes must be decontaminated by scraping before new catalyst is charged. Simple treatments such as steaming are not effective.

Other poisons

Chlorine and other halogens are detrimental and have an effect comparable with that of sulphur. The same concentration limits apply. As with sulphur, the effect of chlorine and chlorides is reversible, and the performance of a catalyst which has been poisoned with feedstock containing 1 ppm of chlorine returns to normal by operation on pure feed. It is possible (see p. 28) that halogens in larger quantities might cause some permanent deactivation of the most active nickel reforming catalysts, such as catalyst 57–1, by a sintering process.

The reforming activity of nickel is also decreased by some metals. Copper and lead should be excluded from feedstocks as, like arsenic, they accumulate on the catalyst and cannot be removed. They are usually removed from naphtha feedstock by the normal desulphurisation process. Concentrations of lead up to 3 ppm can be tolerated for short periods of up to a few days. Silver and vanadium behave in the same way but cadmium, which can be evolved in small amounts from the zinc oxide in desulphurisation catalyst, has no effect.

Catalyst performance

Primary catalysts

The activity of reforming catalysts can be expressed in several ways as, for example, the per cent conversion of inlet methane, or methane content of exit gas (methane slip) at a stated temperature, pressure and throughput. Most frequently the approach to the methane-steam equilibrium is quoted; a temperature difference which is related to catalyst performance.

The minimum value of the methane concentration in the exit gas is defined by chemical equilibrium, which takes account of temperature, pressure, steam ratio, feedstock, etc., as was discussed on pp. 65–72. Thus a decrease in temperature increases the amount of methane which can theoretically be obtained. In practice the methane content of the exit gas at actual temperature T_A is invariably higher than the theoretical minimum so there is a lower 'equilibrium' temperature T_E at which this higher methane content would be at equilibrium. The difference between the actual temperature T_A and the lower temperature T_E at which the methane content would be at equilibrium is termed the approach to equilibrium, ΔT:

$$\Delta T = T_A - T_E.$$

The approach ΔT is always positive when methane or naphtha are being reformed to make synthesis gas (although it can be negative when town gases rich in methane are being made from naphtha), and low values of ΔT (for example below 20°C) correspond to high catalyst activity. The approach is affected by space velocity (throughput), by steam ratio and, to a lesser extent, by temperature and pressure. To assess ΔT the operating conditions have to be defined accurately, but over the usual range of reforming conditions at temperatures up to 800°C and pressures up to 450 lb/in^2 gauge, catalyst 57-1 in a primary reformer gives an approach of 10–15°C.

Similar considerations apply to the naphtha-reforming catalyst 46-1, but in this case there is the additional requirement that substantially complete reforming of the naphtha feed has to be obtained. The operating conditions are very dependent upon the usual operating variables and the composition of the feedstock, and particularly upon the amount of aromatic compounds the feedstock contains. With a light naphtha feedstock containing 4 per cent benzene and toluene, a reformer operating with catalyst 46-1 at an exit temperature of 750°C will produce gas containing less than 1500 mg of benzene + toluene/m^3. At 800°C and above, the amount becomes insignificant—less than 50 mg/m^3. Determination of the concentration of aromatics in the product gas provides a convenient measure of catalyst activity. The effect of any poisons in the feedstock can be detected by an increase in the benzene and toluene concentration before any change in methane content of the product gas is noted. This is consistent with laboratory and semi-technical experiments, which have shown that aromatic compounds reform at a slower rate than the aliphatic paraffins.

In addition to activity, another essential requirement of a naphtha-reforming catalyst is that it shall function without forming carbon at low steam ratios—between 3·0 and 4·0. In order to have some margin of safety when operating a full-scale reformer continually at this level, the catalyst must be able to run at an even lower ratio. Catalyst 46-1 was developed to meet this requirement and will operate indefinitely, in carefully controlled semi-technical experiments, at pressures up to 40 atm, at a steam ratio of 1·5, which is approaching the theoretical minimum of 1·2 (fig. 37).

The ICI semi-technical plant used for these experiments (as well as all the development work of reforming catalysts) is shown in fig. 46. It consists of a full-scale reformer tube of 10 cm (4 in) internal diameter, 6·1 m (20 ft) heated length, which can be operated at pressures up to 50 atm, and catalyst temperatures up to 850°C, and with a range of gaseous or liquid feedstocks. This scale of experimentation is essential if reliable results regarding activity, poisoning, and minimum operable steam ratio are to be obtained. Small-scale pressure apparatus using smaller tubes can be used to compare catalyst activities, but because the catalyst particles have to be smaller than those used in full-scale operations, the results are of little value in absolute terms, as for example, in the determination of a minimum steam ratio.

Information on catalyst life can only be obtained from full-scale reformers, and experience has shown that catalyst 46-1 has a life of 1–3 years, or sometimes longer, depending on the severity of the operating conditions. The lower figure applies at high temperatures and pressures (above 800°C and 400 lb/in^2 gauge). Natural-gas catalysts have a considerably longer life than this, though the duration again depends on operating factors. Reforming catalysts rarely need to be discharged for loss of activity due to poisoning, since in most cases this is reversible. The exception is arsenic, and a catalyst poisoned with arsenic must be changed.

The usual reason for a catalyst reaching the end of its useful life is the consequence of a slow, but steady, increase in the resistance to flow through it. This is due to catalyst distintegration or (and) carbon deposition. At the same time the tube becomes hotter. Carbon, unless the tube has been completely blocked, can be removed by steaming the catalyst, at 800°C, for 12–24 hours. A pressure drop due to disintegration cannot be improved by steaming so

FIG. 46. Steam reforming semi-technical plant.

the cause of the obstruction can be identified in this way. Overheated tubes, in the absence of an increased pressure drop, are indicative of poisoning, and this should be apparent in a higher methane slip, an increased approach to the methane equilibrium and (in naphtha reforming) by the presence of aromatics in the product gas. The onset of carbon formation is sometimes revealed by a loss of activity.

Catalyst failures are usually caused because some plant mishap, or unsteady running, has resulted in the catalyst being subjected to a low steam-ratio, to overheating or to poisoning. Operation at steam-ratios below the thermodynamic minimum (fig. 37) for very brief periods of time can cause severe catalyst disintegration as a result of carbon lay-down within the catalyst ring, and this usually makes a complete catalyst change necessary. Catalysts 57–1 and

46–1 have high physical strength and are able, briefly, to withstand partial steam failure. Similarly they can withstand short periods of overheating up to 1000°C, although prolonged periods (longer than one hour, for example) should be avoided as this can cause a decrease in catalyst activity due to some loss in nickel surface area.

Secondary reforming catalysts

The secondary reformer extends the duties of a methane reforming catalyst. Primary gas at 750–800°C containing H_2, CO, CO_2, H_2O, and up to 10 per cent CH_4 (dry analysis) is re-equilibrated at temperatures up to 1000°C in order to reduce the methane content to about 0·2 per cent (see pp. 65–66 and figs. 29, 32 and 38). The temperature is obtained by introducing air, which also provides the nitrogen for ammonia synthesis. The air is mixed with the primary gas, in a zone above the catalyst bed, and a rapid reaction is started. Steam reforming probably occurs in this 'flame' area, and this will tend to lower the temperature there. But a theoretical 'flame' temperature can more easily be calculated if this effect is neglected. It can be as high as 1200°C, depending on air rates, pre-heat temperatures, and so on. The temperature then falls, and the exit temperature is around 1000°C. A secondary catalyst must therefore be able to withstand temperatures of 1200–1300°C without breaking or shrinking. Sometimes a layer of a more refractory catalyst is put on top of the main catalyst to protect it from the 'flame' but this is not necessary with catalyst 54–2, which is satisfactory up to at least 1300°C. As a result of bad mixing between the air and primary gas considerably higher 'flame' temperatures are possible (on occasions temperatures over 1500°C have been reported); these temperatures can only be avoided by careful design of the mixer. The maximum protection to the catalyst is given when the bed is covered with a layer of highly refractory material (such as alumina containing no nickel or other impurities which lower its melting point). This will enable the catalyst to withstand a degree of poor mixing. Poor mixing, however, invariably leads to a high proportion of methane in the exit gas because there is little mixing inside the catalyst bed.

Catalysts for secondary reforming are of the nickel–high alumina–low silica type. The significance of these components has been discussed earlier on pp. 79–81. Retention of shape and physical strength are the main requirements. Because of the high temperature of operation activity is less important, but the catalyst formulation must be such as to retain the maximum nickel surface area and avoid gross sintering.

Kinetics and mechanism

Methane reforming

The catalytic reaction of methane and steam has been studied by a number of workers, notably by the Russians: Bodrov, Apel'baum and Temkin (ref. 57). Akers and Camp (ref. 58) studied the effect of composition on the rate of this reaction, using a nickel on kieselguhr catalyst in an integral reactor at a temperature of 638°C and a pressure of 1 atm. They found that the reaction was first-order with respect to methane, that both CO and CO_2 were primary products, and that the water gas shift reaction was either absent or was very slow. They suggested that the chemisorption of CH_4, or the decomposition of CH_4 to CH_2 radicals and H_2, was the rate-controlling step, and an activation energy of 15 800 Btu/lb mole (9 kcal/mole) was derived.

The Russian workers (ref. 57) used a circulation flow method to study the kinetics. Their measurements were made at 800–900°C and 1 atm pressure, and nickel foil was employed to obviate any pore-diffusion limitation. They reached six main conclusions.

(a) The decomposition of CH_4 and the reaction of CH_4 with steam or CO_2 was negligible in the absence of catalyst.

(b) Water gas shift equilibrium was always established.

(c) The data could be satisfactorily described by the equation

$$\text{rate} = k \frac{P_{CH_4}}{1 + a(P_{H_2O}/P_{H_2}) + bP_{CO}},$$

where at 800°C $a = 0.5$ and $b = 2.0$ atm^{-1},
at 900°C $a = 0.2$ and $b = 0$ atm^{-1}.
and k is the rate constant.

(d) The activity of the catalyst was affected by its previous history.

(e) An activation energy of 31 kcal/mole for the range 800–900°C was calculated.

(f) The decomposition of CH_4 to carbon on the nickel surface was many times slower than the reaction of CH_4 with steam, making any mechanism based on the initial decomposition of CH_4 to carbon and hydrogen improbable.

The decomposition of CH_4 on nickel surfaces (ref. 59) and the exchange of CH_4 with deuterium on nickel film (ref. 60) have been studied, and chemisorbed CH_2 and CH_3 radicals were found. The activation energy of 31 kcal/mole coincides with that of the exchange of CH_4 with deuterium, and these considerations, together with the experimental rate expression, suggest a mechanism in which the initial rate-determining step is the chemisorption of CH_4 to give chemisorbed CH_2 radicals and H_2. The activity of the catalyst depends on its previous history, the factors involved probably being oxidation and re-reduction of the surface under high steam partial pressures, and the removal of surface carbon formed by the decomposition of chemisorbed CH_2 radicals.

The same Russian workers (ref. 61) have recently repeated their studies over nickel catalysts (Ni on alpha alumina). A simple kinetic equation is sufficiently accurate for this case.

$$\text{rate} = k\, P_{CH_4}$$

Activation energies of 18·3 kcal/mole at 800–900°C, and 24 kcal at 700–800°C were found. These are substantially smaller than on nickel foil, and this is attributed to diffusion limiting the overall rate of reaction. The effective thickness of the working layers was calculated, being 0·04 mm for 1·2 mm particles.

Naphtha reforming

Although most published work concerns methane reforming there are several reasonable schemes which give some account of the naphtha-reforming process. Three different mechanisms can be postulated.

(i) Initial decomposition of the hydrocarbons to carbon and hydrogen, followed by reaction of carbon with steam.
(ii) Stepwise breakdown of the hydrocarbons, and direct reaction of steam with hydrocarbon fragments on the catalyst surface.
(iii) Direct reaction of hydrocarbons with steam, giving oxygen-containing intermediates, possibly in the form of surface compounds on the catalyst.

It is not clear whether CO_2 or CO is the first-formed oxide of carbon, or whether both are formed simultaneously, but it is generally accepted that the following reactions between H_2, CO, CO_2, H_2O, and CH_4 are catalysed at some stage of the process, and tend to effect complete equilibrium among all the components of the product gas.

$$CO + H_2O \rightleftharpoons CO_2 + H_2$$

$$CH_4 + H_2O \rightleftharpoons CO + 3H_2$$

$$CH_4 + 2H_2O \rightleftharpoons CO_2 + 4H_2$$

$$CH_4 + CO_2 \rightleftharpoons 2CO + 2H_2$$

The rate of the reaction of steam with carbon deposited on the surface of a steam-reforming catalyst has been compared with the rate of the hydrocarbon–steam reaction over the same supported nickel catalyst. The results suggest that the reaction of steam with massive carbon is too slow to support a mechanism involving the initial decomposition of hydrocarbons to carbon and hydrogen. There is little doubt that the process takes place in a series of simpler, intermediate steps, which probably form a network of parallel and consecutive reactions.

In ICI, experiments have been made in which paraffins were reformed in a small catalyst bed at high space velocities, over a wide range of temperatures (500–800°C), and at atmospheric pressure. The results suggested that the steam reforming reaction is a combination of homogeneous thermal cracking, and heterogeneous catalytic reaction, of hydrocarbons with steam. At high temperatures—about 700–750°C—the activation energy of the paraffin–steam reaction was found to be almost identical with published values for homogeneous paraffin thermal cracking. At lower temperatures—below 650°C—the reaction had a smaller activation energy, typical of catalysed reactions, and a number of different paraffins showed qualitatively similar behaviour.

Subsequent investigation of the catalytic reaction over nickel catalysts at temperatures of about 600°C showed that the paraffins are first cracked catalytically, mainly to olefines and methane. These primary products react with steam, yielding hydrogen and oxides of carbon. The process thus produces olefines and methane selectively at short contact times, and hydrogen and carbon oxides at long contact times.

The conclusions from all these experiments can be used as a basis for a discussion of the behaviour of paraffins in the steam reforming process. Fig. 47 presents a simple scheme for the reactions between paraffins and steam, and there are three major stages.

(i) Catalytic cracking and dehydrogenation and, at high temperatures,

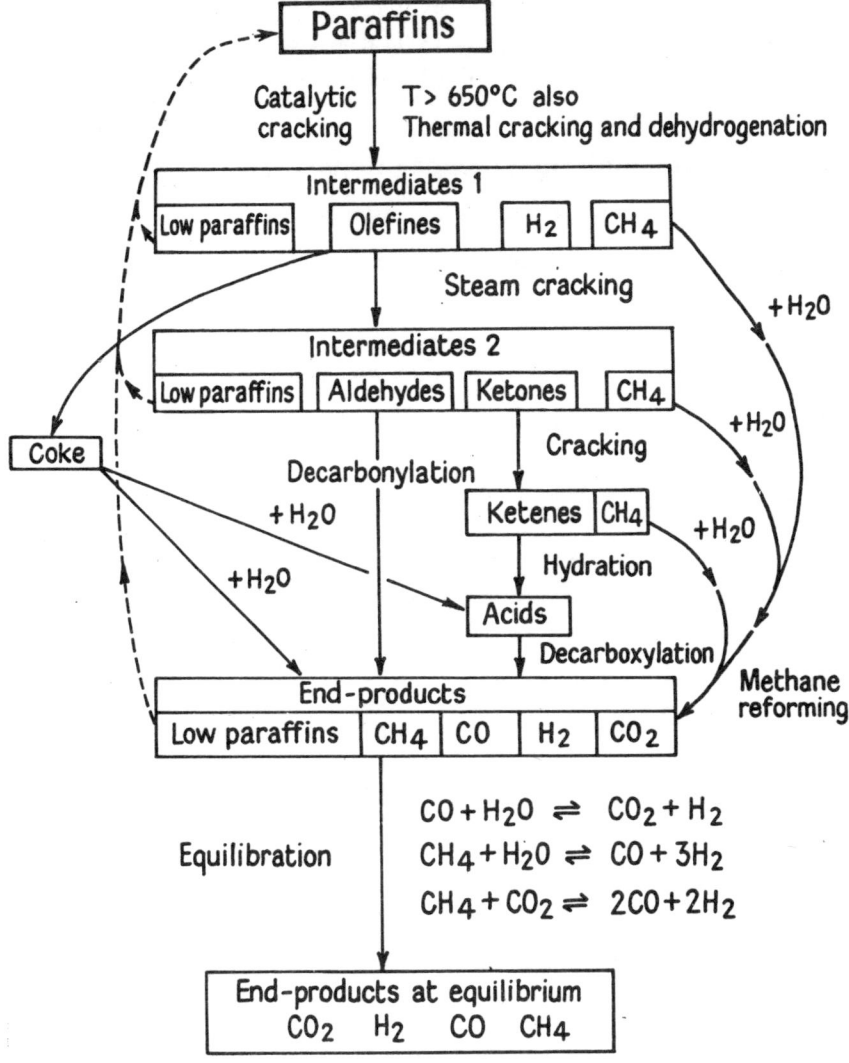

FIG. 47. Steam reforming of paraffins.

thermal cracking—leading mainly to olefines of low molecular weight, methane, and some hydrogen.
(ii) Reaction of primary intermediates with steam—leading to hydrogen and carbon oxides.
(iii) An equilibration stage between H_2, steam, CO_2, and CO and CH_4.

It seems unlikely that stage (ii) takes place in simple steps, but no intermediates have been detected. The evaluation of a 'virtual mechanism' (Chapter 1) has led to the postulated routes shown in the diagram (postulated intermediates are shown in hatched boxes). It is postulated that carbon, if formed, comes from olefines by a scheme involving successive polymerisation and dehydrogenation reactions. This can explain why acidic catalysts catalyse the formation of 'coke' from olefines via aromatic systems.

Calculation of equilibrium compositions

The steam-reforming reaction can be written as

$$CH_{2\alpha} + xH_2O \rightarrow aCO + bCO_2 + (1-a-b)CH_4 + (x-a-2b)H_2O + (3a+4b+\alpha-2)H_2,$$

assuming a mass balance and no carbon deposition.

For equilibrium with respect to the water-gas-shift reaction and the methane–

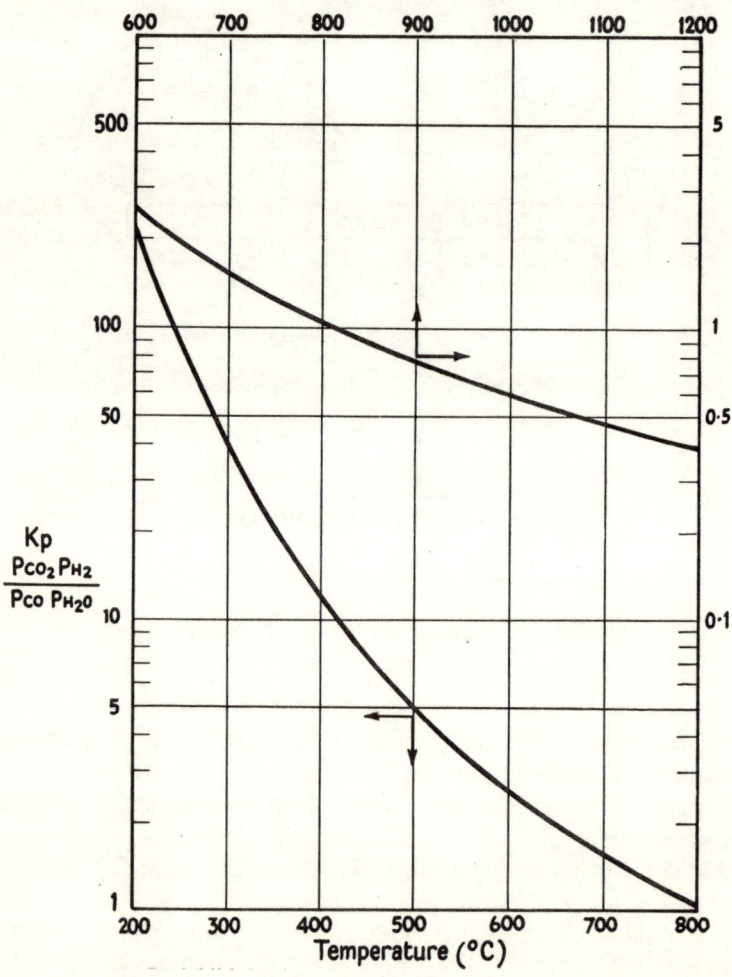

FIG. 48. Water gas shift equilibrium. K_p as a function of temperature.

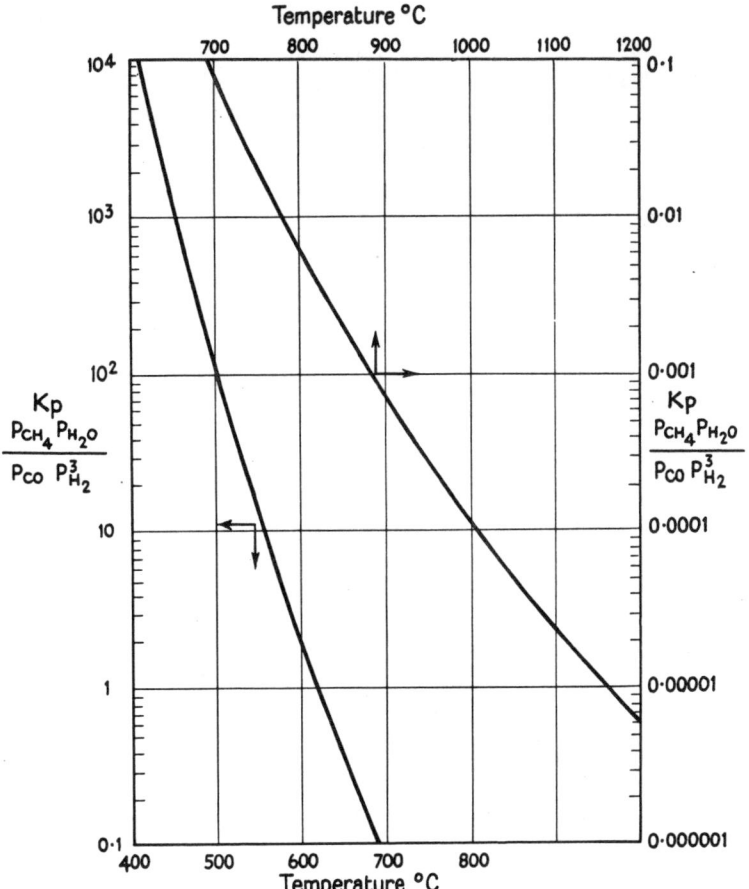

FIG. 49. Methane − steam equilibrium. K_p as a function of temperature.

steam reaction, two equations can be derived (P is the pressure in atm).

$$K^T_{\text{W.G.S.}} = \frac{P_{CO_2} P_{H_2}}{P_{CO} P_{H_2O}} = \frac{b(3a+4b+\alpha-2)}{a(x-a-2b)}, \tag{1}$$

$$K^T_{M/S} = \frac{P_{CO} P^3_{H_2}}{P_{CH_4} P_{H_2O}} = \frac{a(3a+4b+\alpha-2)^3}{(1-a-b)(x-a-2b)} \times \frac{P^2}{(x+2a+2b+\alpha-1)^2} \tag{2}$$

If $K^T_{\text{W.G.S.}}$ and $K^T_{M/S}$ are known as a function of temperature, for a given temperature, pressure, feed composition (value of 2α), and steam ratio (x) there are two unknowns, a and b, and two simultaneous equations. The solution of these equations for a and b is a trial-and-error process, and is best carried out on a computer (a solution is described in the appendix).

Graphs of $K^T_{\text{W.G.S}}$ and $K^T_{M/S}$ as functions of temperatures are given (figs. 48 and 49) and there are tables of values in the appendix.

To find the minimum operable steam ratio without carbon deposition, a further restriction—that one of the carbon forming reactions shall be at equilibrium—is imposed. This, together with the condition of zero carbon deposition, fixed the boundary conditions required. It is immaterial which carbon-forming reaction (a), (b), or (c) is chosen because any of these reactions can be derived from any other in combination with the methane–steam and water-gas-shift equilibria.

(a) $2CO \rightleftharpoons C + CO_2$
(b) $CO + H_2 \rightleftharpoons C + H_2O$
(c) $CH_4 \rightleftharpoons C + 2H_2$

If (b) is chosen the further restraint is

$$K_{(b)}^T = \frac{P_{H_2O}}{P_{CO} P_{H_2}} = \frac{(x-a-2b)}{a(3a+4b+\alpha-2)} \times \frac{(x+2a+2b+\alpha-1)}{P}. \qquad (3)$$

For a given temperature, pressure, and feed there are now three equations and three variables, a, b, and x (minimum steam ratio). Hence x can be found.

For secondary reforming and CO_2 reforming the calculations are similar, except that the overall reaction and mass-balance equation has to be modified to take account of the inlet air or CO_2, with consequent modification to equations (1) and (2).

CHAPTER 6

Removal of carbon monoxide

AMMONIA synthesis-gas is normally produced from feedstocks containing carbon. Oxides of carbon, which de-activate the ammonia synthesis catalyst (Chapter 7), must therefore be removed before the synthesis-gas can be utilised. Most modern plants steam-reform hydrocarbons, convert carbon monoxide to carbon dioxide in two stages of shift conversion, absorb CO_2 in a scrubber and finally purify the synthesis-gas by methanating residual CO and CO_2 down to trace levels. Alternative techniques, such as CO absorption with copper-liquor or purification by low-temperature distillation, usually have higher running costs and sometimes higher capital costs than the catalytic system but can be favoured by special circumstances at individual factories.

High-temperature shift conversion, reacting carbon monoxide with steam to form carbon dioxide and water, increases the efficiency of hydrogen utilisation and, consequently, is employed on almost all ammonia plants. Low-temperature shift conversion is a relatively new process that requires pure gas streams and modern catalyst technology. It offers a small gain in hydrogen but its main advantage is to lower the carbon monoxide content to a level that avoids costly absorption equipment. Methanation (reacting CO and CO_2 with hydrogen to form methane) is not a new process but its application to ammonia synthesis-gas production was rarely possible before the advent of low-temperature shift catalysts.

The water-gas shift (CO conversion) reaction

Carbon monoxide and steam can, with suitable catalysts, be converted to carbon dioxide and hydrogen:

$$CO + H_2O \rightleftharpoons CO_2 + H_2. \qquad \Delta H = -9\cdot 84 \text{ kcal} \qquad (1)$$

This exothermic reaction is reversible, the extent to which maximum conversion is limited by equilibrium being governed by the thermodynamic considerations set out at the beginning of Chapter 1. The equilibrium is independent of pressure and the table of equilibrium constants given in the appendix shows how the equilibrium carbon monoxide concentration increases with temperature, favouring high conversions at low temperatures.

Catalysts for redox reactions (fig. 2), such as reaction (1), must have both the active and selective properties, characteristic of a mild hydrogenation function, to achieve adequate reaction rates without also forming methane. Appropriate catalysts can therefore be expected among the Group 1b metals (p. 12), Group

VIII oxides and Group VIII sulphides (fig. 4). The further restriction, that the catalyst must be stable in the reaction gas, limits the choice to Cu metal, Fe_3O_4, and FeS. In addition, a sulphided form of cobalt molybdate, to a limited degree, has suitable properties.

In practical terms Fe_3O_4 is the base of a good general catalyst that has advantages of cheapness, stability and an ability to withstand considerable quantities of impurities without being poisoned. In gases containing relatively large quantities of sulphur, FeS is the best available catalyst, and a further advantage of Fe_3O_4 is that it forms FeS in a sulphur atmosphere, allowing adequate CO conversion even under extreme conditions. The disadvantage of Fe_3O_4 is that relatively high temperatures, typically over 350°C, are required before the catalyst has sufficient activity for most commercial applications. Consequently the degree of conversion obtainable on Fe_3O_4 (that is, high temperature CO shift catalysts) is limited and is inadequate for most modern plants.

The activity of copper metal is greatly superior to Fe_3O_4, with the consequence that it has practical applications even below 200°C. Relatively high cost and susceptibility to poisons are minor disadvantages, outweighed by the high conversions that are obtained at temperatures below 250°C. Copper catalysts, in the form of low temperature shift catalysts, consist of very finely divided particles. Operation, therefore, is restricted to a limited temperature range (p. 23, structural properties) above which activity is rapidly lost. For this reason, as well as catalyst cost, conversion in the LTS converter is normally restricted, in order to limit the temperature rise, and the converter is preceded by conventional high-temperature shift conversion.

High-temperature shift (CO conversion) catalysts

The CO shift conversion reaction has been used for producing hydrogen from carbon monoxide for over 40 years. The basic catalyst was Fe_3O_4, but the addition of Cr_2O_3 to the catalyst was found to inhibit crystal growth during operation (pp. 24–25), greatly increasing the useful life of the catalyst, and commerical high temperature CO shift catalysts contain mixtures of Fe_3O_4 and Cr_2O_3. ICI catalyst 15–2, first produced in 1936, was a catalyst of this type and it has been employed in processes operating from atmospheric pressure up to 30 atmospheres, in gases with a wide range of carbon monoxide contents, rising above 60 per cent CO, and with both sulphur-free gases and mixtures containing more than 50 000 ppm of H_2S.

ICI experience
Certain lessons were learnt from the experience gained with catalyst 15–2. The catalyst was exceedingly robust and each pellet could normally withstand a pressure of over 400 lb wt. This was associated with a high bulk-density of 1·35 kg/l, and undoubtedly contributed much to the long and reliable operation which was obtained, periods of service sometimes exceeding 12 years on low-pressure plants. Break-up of the catalyst was unknown, even when it was subjected to overheating during the regeneration cycles encountered in certain gas-forming processes. On the other hand it was realised that much of the

catalytic material inside the pellet was not available for reaction because diffusivity was relatively low (see Chapters 2 and 3).

The method of manufacture of this catalyst made it inevitable that about 1 per cent of sulphates was present. The sulphates were slowly reduced and during the early life of the catalyst H_2S was evolved. Since H_2S is a serious poison for LT CO-shift catalyst the presence of sulphate in the HT shift catalyst is undesirable. For some plants producing town's gas H_2S is also clearly disadvantageous. This experience has led to changes in catalyst formulation.

Other catalysts

Other catalyst manufacturers have generally experienced problems similar to those encountered by ICI and have adopted a variety of techniques to overcome them. A number of catalysts have been made in the form of lumps of unpelleted material which, in addition to lowering the price of the catalyst, partly solves the problems associated with diffusivity but only at the expense of catalyst strength. Before use, HT shift catalysts are reduced, removing some oxygen, and this lowers their physical strength. During use, break-down of the catalyst causes some dust to form, and consequently the pressure-drop across the converter increases steadily with age. The rate of increase in pressure-drop is thus largely determined by the strength of the catalyst particles. In recent years the move has been towards pelleted catalysts, and pellets in the form of Raschig rings have also appeared on the market. To combat diffusivity limitations, and to meet commercial pressure on production costs, the bulk densities have remained relatively low, usually below 1·1 kg/l. Although initial pressure-drop for freshly charged catalyst is now low, the strength of the catalyst after reduction is not sufficient to prevent a steady increase in pressure-drop due to dust-formation. There has also been much research on production techniques that extend the operating range of the catalyst to lower temperatures, and it is possible to obtain catalysts which will start operation at 280°C and which can, therefore, operate under conditions where the reaction equilibrium is more favourable. Unfortunately, this low-temperature activity is only present when the catalysts are new and it is readily destroyed by heating to slightly higher temperatures. The search for a form of Fe_3O_4 which will operate at lower temperatures has lost much of its impetus since the development of copper catalysts which can operate with exit temperatures below 250°C. The problem of residual sulphates in HT shift catalyst appears to affect all manufacturers and frequently has been cited as a cause of failure of low-temperature shift-catalysts.

Present requirements

The structural requirements of an ideal catalyst are shown in Chapter 2 (fig. 5). ICI have regarded strong physical properties as the prerequisite of HT shift catalyst because the useful life of these catalysts is long and is so frequently terminated by pressure-drop considerations rather than loss of activity. Modifications to the manufacturing techniques have largely prevented the inclusion of residual sulphate, and catalysts can be manufactured containing less than 0·1 per cent sulphur. These catalysts can be reduced without any precautions being taken to prevent the sulphur reaching a town's gas holder or an LT shift converter. It is preferable, however, not to use the gas during the time when sulphur evolution exceeds 1 ppm, which is for less than 24 hours. This has led to the replacement of catalyst 15-2 with catalyst 15-4, which has the strength and activity of the earlier catalyst but is made with a low sulphur-content. To surmount diffusivity limitations the catalyst is now also manufactured in the small pellet size (5·4 mm dia. × 3·6 mm high) known as catalyst 15-5, an application of the techniques of Chapter 2 where fig. 5 illustrates how correct flow-properties permit full utilisation of catalyst activity. Because the catalytic material is employed in this efficient manner the volume of catalyst 15-5 required for any specified duty is considerably lower than that of other catalysts. The bulk density of catalysts 15-4 and 15-5 is 1·35 kg/l and remains unchanged from catalyst 15-2, to ensure that strong pellet bonding maintains the correct fluid-flow properties throughout the long, useful life.

Reactions with HT shift catalysts

The water-gas shift reaction

The exothermic reaction between steam and carbon monoxide to yield hydrogen and carbon dioxide was described near the beginning of this chapter (reaction (1)).

High conversions are favoured by low temperatures but the rate of reaction increases as the temperature is raised. The rate of conversion, consequently, increases with temperature but decreases when the gas composition is close to equilibrium.

Conversion of 1 per cent CO, in the wet gas, raises the temperature approximately 10°C so there is an increase in temperature across a shift converter. The exit-gas temperature is normally determined by making appropriate considerations about the equilibrium concentration of CO demanded by design performance, and the inlet temperature is defined by the temperature rise across the converter. Under normal operating conditions the temperature rises progressively through the converter, but the rate of conversion reaches a maximum and then falls again until, near the exit, it is limited by the reaction's close approach to equilibrium.

Catalyst requirements can be lowered if the catalyst is divided into more than one bed, the beds being connected in series, using interbed cooling. This lowers the temperature-rise across each bed, which allows the rate of conversion in each to be increased. It also allows the average operating temperature to be increased with consequent increase in the average rate of conversion. Interbed cooling can be provided by heat exchangers, when the heat is required for other purposes, or by adding quench-water or steam. The addition of quench has the added advantage of increasing the steam concentration, which reduces the equilibrium concentration of CO in subsequent beds.

High concentrations of carbon monoxide, such as arise from partial-oxidation plants, require special consideration. The catalyst is usually divided into three beds, with inter-cooling, and this arrangement allows the reaction to be completed within the operating temperature range of the catalyst. Addition of quench water allows the carbon monoxide concentration in the exit gas to be lowered to a satisfactory level, even in the presence of a high concentration of carbon dioxide. Problems can arise, however, during start-up and shut-down, if certain requirements are overlooked. The gas at the exit from the first catalyst bed will, effectively, reach equilibrium whenever throughput is low. To ensure a reliable operation the inlet and exit temperature must be inside the working-range of the catalyst. The temperature rise across the first bed and consequently the conversion, must therefore be limited. This can be achieved by increasing the ratio of steam to dry gas until the adiabatic temperature rise across the converter is controlled in the working range, and until the maximum temperature of the gas is controlled, by equilibrium, below the maximum working temperature of the catalyst.

Catalyst reduction

The catalyst is reduced in process gas, a mixture of carbon monoxide, hydrogen, carbon dioxide, and steam. Details of the methods used are described in Chapter 9. Reducing conditions must be chosen which permit Fe_2O_3 to reduce to Fe_3O_4 without the possibility of further reduction to metallic Fe.

The required reduction reactions are

$$3Fe_2O_3 + H_2 \rightarrow 2Fe_3O_4 + H_2O \qquad \Delta H = -2.30 \text{ kcal} \qquad (2)$$

$$3Fe_2O_3 + CO \rightarrow 2Fe_3O_4 + CO_2 \qquad \Delta H = -12.14 \text{ kcal} \qquad (3)$$

The equilibrium between the phases Fe_2O_3 and Fe_3O_4 is determined by the ratio of H_2/H_2O or of CO/CO_2. Fe_3O_4 is the stable phase under normal conditions and the reactions can only be reversed by employing H_2O or CO_2 containing less than 500 ppm of H_2 or CO.

Over-reduction of the catalyst must be avoided because it results in metallic Fe which can catalyse the highly exothermic reaction between carbon monoxide and hydrogen to form methane, and this can damage both catalyst and converter

$$CO + 3H_2 \rightarrow CH_4 + H_2O \qquad \Delta H = -49.27 \text{ kcal} \qquad (4)$$

Metallic iron can probably also catalyse the disproportionation of carbon monoxide and cause carbon to be formed

$$2CO \rightarrow C + CO_2 \qquad \Delta H = -41.22 \text{ kcal} \qquad (5)$$

The reduction reactions to be avoided are

$$Fe_3O_4 + 4H_2 \rightarrow 3Fe + 4H_2O \qquad \Delta H = +35.81 \text{ kcal} \qquad (6)$$

$$Fe_3O_4 + 4CO \rightarrow 3Fe + 4CO_2 \qquad \Delta H = -3.54 \text{ kcal} \qquad (7)$$

Reduction with pure hydrogen, or with a hydrogen/nitrogen mixture, is not recommended because metallic iron is readily formed. There are several reported incidents which resulted in a high temperature rise when process gas was added after the catalyst had been reduced in hydrogen. Over-reduction is prevented by addition of steam in the hydrogen. Equilibrium data indicate that 10 per cent steam in hydrogen at 400°C, or 17 per cent steam at 550°C, is sufficient to prevent this reaction.

The equilibrium for over-reduction with carbon monoxide favours the reduction reaction even when carbon dioxide is the major constituent. The carbon monoxide to carbon dioxide ratio in typical dry process gases is frequently, at the converter inlet, above the thermodynamic limit for reduction of the catalyst to iron. The steam in the process gas ensures that the catalyst is not reduced to iron. It also promotes the water-gas shift reaction and lowers the CO/CO_2 ratio below the range where the catalyst might be reduced.

Reactions of sulphur compounds

When the process gas contains sulphur compounds, such as H_2S or COS, in concentrations below about 200 ppm they have no effect on the normal activity of the catalyst. The catalyst can, however, absorb the sulphur compounds, and it releases them slowly. This effect can be important when the high temperature shift catalyst is being operated in conjunction with a low temperature shift catalyst. For example, when primary and secondary reforming catalysts are reduced, H_2S is frequently formed, and during this operation the LT shift converter is normally isolated. If the sulphur is passed to the HT shift converter it will be stored there and, after the LT shift converter is reconnected, much of the H_2S will ultimately pass into it. If the sulphur concentration in the inlet gas to the HT shift converter exceeds about 200 ppm, sulphur storage takes place via an alternative mechanism (see paragraph on the formation of FeS). The overall effect of sulphur storage in HT shift catalysts plays an important role in LT shift catalyst poisoning, and in this connection it is advantageous to avoid passing sulphur compounds through HT shift converters.

A small proportion of sulphate is normally present in HT shift catalysts because some insoluble sulphates form during its preparation. During use, in the presence of hydrogen and steam, sulphates are slowly reduced and H_2S is evolved. Frequently catalysts contain between 0·5 and 1·0 per cent SO_3 as residual sulphates, and old-fashioned production methods make catalysts which require three to four days or longer before the sulphur levels are safe for LT shift converter operation. By using special preparative techniques, such as are employed with catalysts 15–4/15–5, it is possible to lower the sulphate content of the catalyst, and also to modify the constitution of the residual sulphates. In catalyst 15–4 and 15–5 there is about 0·1 per cent SO_3 in the form of sulphates which are difficult to remove, and up to approximately the same quantity in a rapidly reduceable form. When catalysts 15–4 and 15–5 are reduced the H_2S evolved from new catalyst is frequently below 1 ppm. When it is higher, the excess is rapidly removed, and the level falls below 1 ppm during the first day's operation. The maximum quantity of sulphur which can be evolved from catalyst 15–5, after the first 24 hours on line, is 0·1 per cent SO_3—the usual figure being below 0·04 per cent. This, coupled with the high activity per unit weight of the catalyst, limits the amount of sulphur that can reach the LT shift catalyst, to an acceptable level.

FeS is a satisfactory catalyst for the water-gas shift reaction, and commercial charges are frequently used in this state. The activity of FeS is approximately half that of Fe_3O_4 and in these circumstances the calculated catalyst volume must be increased by a factor of two. The activity of FeS is unaffected by sulphur compounds in the process gas, and the catalyst can be used over a wide range of gas compositions. Catalyst 15–2 has been operated on plants where H_2S was about 50 000 ppm. In the laboratory, where more sensitive tests can be used, a slightly different picture emerges, as follows. A desulphided catalyst operated in gas containing less than 1 ppm sulphur has higher activity than freshly reduced catalyst. Addition of 50 ppm H_2S to the inlet gas causes the activity to fall to a normal level, and during the time the activity is falling the exit gas contains less than 50 ppm H_2S. Addition of 100 ppm H_2S causes activity to fall slightly below that of normal freshly-reduced catalyst. Again the sulphur in the exit gas does not reach 100 ppm during the activity change. The effects, which are reversible, are an indication that Fe_3O_4 absorbs sulphur even at low sulphur concentrations, that the quantity of sulphur absorbed is related to the partial pressure of sulphur, and that absorbed sulphur acts as a mild poison.

The theoretical conditions for the formation of FeS from Fe_3O_4 are shown by the equation

$$Fe_3O_4 + 3H_2S + H_2 \rightarrow 3FeS + 4H_2O \quad \Delta H = -17\cdot91 \text{ kcal} \quad (8)$$

Appropriate values of the equilibrium constant are shown in fig. 50.

Temperature (°C)	$\dfrac{H_2S^3 \times H_2}{H_2O^4}$
300	$3\cdot22 \times 10^{-10}$
350	$8\cdot70 \times 10^{-10}$
400	$2\cdot11 \times 10^{-9}$
450	$4\cdot58 \times 10^{-9}$
500	$9\cdot32 \times 10^{-9}$
550	$1\cdot82 \times 10^{-8}$

FIG. 50. Equilibrium constant for Fe_3O_4/FeS reaction.

The sulphiding reaction is reversible, and when the H_2S level falls below a limiting value the catalyst reverts to Fe_3O_4. The formation of FeS from Fe_3O_4, and the reverse reaction, both apply strains to the catalyst causing it to weaken. Catalyst 15–2 has been operated under conditions where sulphiding-desulphiding cycles were made frequently, and the catalyst withstood strains of this operation for a number of years. This may be because the catalyst 15–2, like catalysts 15–4 and 15–5, has a high pellet-strength associated with the high pellet-density. In general terms, however, it is preferable to operate HT shift catalyst as either Fe_3O_4 or FeS, and conditions where frequent sulphiding cycles can occur should be avoided. The conditions to avoid are those where the sulphur levels are about 250–300 ppm H_2S or where they can oscillate both below and above these limits.

The sulphiding reaction can be important during the start-up of plants, where other catalysts are employed in front of the HT shift converter. This is because sulphur, released during the start-up of up-stream catalysts, from their initial reduction or following poisoning, can sometimes reach concentrations which will react with HT shift catalysts. This delays final removal of sulphur from the system, and much of the sulphur may eventually reach the LT shift converter. The advantages of preventing sulphur reaching the HT shift converter are outlined in the first part of this section dealing with sulphur absorption, and they can be re-emphasised in relation to the formation of FeS.

Carbon oxy-sulphide, COS, which is present in some process gases, reacts readily on HT shift catalysts to form H_2S.

$$COS + H_2O \rightarrow CO_2 + H_2S \quad \Delta H = -8.27 \text{ kcal} \quad (9)$$

Equilibrium for this reaction favours the formation of H_2S, as is shown in fig. 51. The reaction is almost complete under normal HT shift reaction conditions.

Temperature (°C)	$\dfrac{COS \times H_2O}{H_2S \times CO_2}$
300	8.35×10^{-4}
350	1.52×10^{-3}
400	2.53×10^{-3}
450	3.55×10^{-3}
500	6.04×10^{-3}
550	8.00×10^{-3}

FIG. 51. Equilibrium constant for COS/H_2S reaction.

Carbon disulphide, CS_2, is expected to undergo a similar reaction to form COS, and the equilibrium constants for this reaction are of the same magnitude as those quoted above. The COS produced by the reaction should, in turn, form CO_2 and H_2S.

Catalyst poisoning

High-temperature shift-conversion catalysts do not normally suffer from troubles due to poisoning. Reforming catalysts, and low temperature shift catalysts, however, are both subject to damage by poisoning and consequently care is taken to ensure that the gases involved in the production of ammonia synthesis gas are normally free of poisons. Large quantities of halogens would deactivate high temperature shift conversion catalysts, but they are not encountered during normal plant operation.

Certain plants, in particular those following cyclic reformers, operate with feed-gases containing traces of acetylene and nitric oxide. These lead to the formation of a gum, containing a high proportion of carbon, which deposits on the catalyst and prevents access of gas to the catalytic surface. Most plants avoid this problem by using a separate bed of a guard catalyst, which can be regenerated. Partial oxidation units sometimes produce a feed gas containing carbon particles which can block pores in the HT shift catalyst.

Regeneration of catalysts blocked with gum or carbon is possible providing that the physical structure of the catalyst pellets is not damaged during 'carbon' formation. Some charges of ICI catalysts 15–2/4 have been regenerated successfully at least four times during their lifetimes. The regeneration consists essentially in treating the catalyst with steam, containing 1–2 per cent oxygen from air, at about 450°C. Some care is needed to prevent the catalyst being overheated, and the temptation to increase the oxygen content during the later part of the regeneration should be avoided. This is because gas distribution through a blocked converter is frequently poor, and regions of unregenerated catalyst may remain after the bulk of the catalyst has been regenerated. Regeneration should continue until the carbon content of the exit gas falls to zero. At this time much of the catalyst will still be in a reduced condition. Special precautions, considered in the next section, are required if the catalyst has been operated in the presence of H_2S.

Reactions with oxygen

The reduced catalyst can be oxidised with oxygen.

$$4Fe_3O_4 + O_2 = 6Fe_2O_3 \quad \Delta H = -111 \cdot 0 \text{ kcal} \qquad (10)$$

When HT shift catalysts have finished their useful lives they must be discharged, and the procedures described in Chapter 9 take into account the exothermic nature of the catalyst oxidation. Under some circumstances, such as a catalyst examination when no inert gas is available, the catalyst requires controlled oxidation with a limited supply of oxygen as described in Chapter 9.

When the catalyst has been operated in the sulphided state (that is to say, part of the catalyst is in the form FeS) it would be damaged by allowing it to react with oxygen

$$6FeS + 13 \cdot 5O_2 = 2Fe_2(SO_4)_3 + Fe_2O_3 \quad \Delta H \text{ approx. } -1344 \text{ kcal} \qquad (11)$$

Following this reaction the catalyst loses activity. The cause is not certain and it could be either that the ferric sulphate cannot be re-reduced to Fe_3O_4, or that the exothermic reaction lowers the surface area of the catalyst. To avoid this particular difficulty the catalyst should be steamed until the FeS has been converted to Fe_3O_4, and it is normally satisfactory to oxidise the catalyst once the level of H_2S in the exit steam is below 80 ppm.

The problem of a catalyst both containing FeS and also blocked with carbon or gums requires special attention. The catalyst can be steamed until the steam contains less than 80 ppm H_2S, but this is no indication that all FeS has reacted with the steam because the steam may not be able to penetrate all regions of the catalyst. To remove carbon, however, some oxygen must be added, and this should be limited to about 0·5 per cent, so that the oxygen reacts with the carbon and allows oxygen-free steam to diffuse slowly into the FeS.

Pellet dimensions for HT shift catalysts

Because users of HT shift catalysts ask for a long life, high pellet strength is an essential requirement, and there is a close relationship between strength and pellet density. Catalyst activity, per unit volume of catalyst, is also related to pellet density. At the extremes, low pellet density indicates too little catalytic material, while high density indicates insufficient voidage for gas to reach the active material. Over a considerable range of pellet densities, however, the activity per unit volume of catalyst is almost constant, and the final choice of pellet density, inside this range, is a balance between strength and cost. ICI, by choosing a high density of 1·35, have ensured adequate strength.

The operation of a converter charged with catalyst is related to catalyst activity and gas-flow properties. These are factors which should determine the dimensions of the catalyst pellets. The internal properties of pellets is the subject of Chapter 2, and Chapter 3 discusses how pellets are utilised in the converter. The effect of pellet-size on internal diffusion (and hence upon the availability of catalytic material), and the effect on gas-flow properties (and hence upon both pressure-drop and gas-distribution), and the effect on strength (and hence upon lifetime), can all be calculated. There are optimum pellet sizes for particular catalytic duties, and, in the two following sections, the sizes of HT shift catalyst pellets are discussed. The premises made in this chapter are somewhat simplified, a more rigorous treatment being given in Chapter 3, but the results are valid.

Effect of particle size on catalyst activity

It can be shown, using the techniques described in Chapter 3, that when a reaction is highly diffusion-limited, only the outer surface of the catalyst pellet contributes towards the reaction. The useful part of the catalyst is proportional to the outer surface of the pellet (S) and the thickness (t) of the useful region. The outer surface of a cylindrical pellet can be expressed in terms of its radius (r) and its height (ar).

$$S = 2\pi r^2(1+a).$$

At the same time the mass (M) of pellets of density d is

$$M = \pi a d r^3.$$

The usefulness of the catalyst is proportional to the outer surface available in a given mass. Thus,

$$\frac{S}{M} = \frac{2(1+a)}{dar}.$$

Hence, when the ratio of height to radius (a) is fixed, and the pellet density is constant, the usefulness of the catalytic material is inversely proportional to the radius of the pellet. This applies whenever the reaction is strongly limited by diffusion, that is, when t is significantly less than r.

Measurements of CO conversion have been made on catalyst material of the 15–4/15–5 type of three different pellet-sizes under standard laboratory conditions, and the results show that all three catalysts are strongly diffusion-limited under normal reaction conditions, even at atmospheric pressure. This is illustrated in fig. 52.

This effect is linked with the high density of catalyst 15–4/5, because the effective thickness of the outer catalyst surface decreases as pellet density increases. At the density of these catalysts, diffusion continues to limit the reaction, even on pellets 3·6 mm high. There are clear advantages, from the point of view of catalyst activity, to be gained from using 5·4 × 3·6 mm pellets. It was shown in Chapter 3, and Chapter 2, fig. 7, that the effects due to diffusion become more pronounced when pressure is increased, and hence there is a

Size of pellets		Per cent conversion at reaction temperature		
diameter (mm)	height (mm)	360°C	400°C	440°C
8·5	11·4	17·7%	26·2%	35·2%
8·5	5·6	20·6%	30·0%	40·2%
5·4	3·6	23·4%	36·3%	50·2%

FIG. 52. Variation of performance with pellet size.

strong case for decreasing pellet-size when increased pressures are used. The optimum size of pellet must also be dependent upon the pressure drop through the converter, and this is discussed in the next section.

Effect of pellet size on pressure drop

The pressure drop through a bed of catalyst pellets arises from two sources, namely the restriction to flow through the spaces between virgin pellets, and the further restriction due to these spaces becoming fouled by process dust and broken pieces of catalyst. Naturally, the first factor (together with the geometry of the bed) determines the pressure drop through freshly packed catalyst, and is, for practical purposes, independent of the catalyst composition. The pressure drop depends on the voidage in the bed, and consequently is at minimum when the catalyst is freshly charged, but increases as the bed packs down. These are geometric properties, and pellets with given dimensions of widely different materials pack in an almost identical manner. The size and shape of the pellets each affect the pressure drop. Pellets which have the height proportional to the radius (that is, $h = ar$, and a is constant) show an increase in pressure drop as the radius is decreased. The shape of the pellet, or the ratio between the height and radius (a, the aspect ratio), also affects pressure drop and pressure drop is highest when $h = 1·7r$. Any shape where the height/diameter ratio is greater or smaller than this ratio lowers the pressure drop. Thus either long, thin, pellets or short, fat, pellets are preferable to nearly square ones. The effect of both shape and size is shown in fig. 53, which compares pressure drops in the case of pellets having a height equal to their diameter, with drops occurring when pellets with other ratios are used.

The effect of pellet size is greater at low flow-rates, and fig. 53 shows data for a low mass flow typical of an atmospheric pressure converter, and for a mass flow typical of a steam reforming plant.

Pressure drop for a given gas flow increases at low pressures, and is inversely proportional to pressure. Design catalyst volumes are lower at higher pressures,

Pellet diameter (mm)	11	8·5	9	6	5	5·4	3
Pellet height (mm)	11	11·4	9	6	5	3·6	3
Low gas-flow rates	1	1	1·3	2·4	3·1	3·1	6·4
High gas-flow rates	1	1	1·2	1·8	2·2	2·2	3·7

FIG. 53. Comparison of pressure drops through beds of pellets of different size and shape.

and are roughly inversely proportional to P raised to the power 0·5. Consequently at low pressures large volumes of catalyst are required, and the pressure drop per unit volume is fairly high, but at pressures above about 10 atm the initial pressure drop through HT shift converters (that is, the pressure drop due to the packed catalyst) is usually less important.

The most important factor in determining the usefulness of an HT shift catalyst is not the initial pressure drop through the bed, but the manner in which, due to gradual fragmentation of the catalyst, the pressure-drop increases. Such dust formation arises when the catalyst is in the reduced state, when the physical strength is considerably lower than that of new catalyst. If two catalysts are compared, that which is the stronger before reduction is normally also the stronger after reduction. This is an argument which favours the manufacture of catalyst of high bulk-density, because increasing pelleting pressures produce stronger pellets. For operating periods of up to approximately 12 months, the hardness of the virgin pellets has little effect on pressure drop, but HT shift catalysts can retain their intrinsic activity for much longer periods. The operating cost and lifetime of HT shift catalysts in converters operating close to atmospheric pressure is usually determined by the pressure drop which develops across them. A catalyst with both a high strength and a low initial pressure drop offers an appreciable advantage in this respect, so that a higher initial cost can be justified.

At pressures over about 10 atm, however, a catalyst with an extra-low pressure drop usually has no special value, and its higher cost cannot be justified. But pressure drop cannot be ignored entirely, and the life of a charge of HT shift catalyst is still often terminated because the drop becomes excessive. High pellet strength, therefore, remains a real advantage on all plants. In catalyst 15–5, high strength has been maintained (and the bulk density of this catalyst remains at 1·35 kg/l), but pellet size (5·4 × 3·6 mm) has been adjusted to suit operation at these higher pressures. Pages 106–107 show how catalyst activity can be increased by a decrease in the pellet size, and the volume of catalyst 15–5 required for a specified duty is roughly two thirds that of any alternative catalyst.

The pressure drop is proportional to the cube of the bed height times the square of the space velocity. For a constant bed diameter and gas rate (since the space velocity is inversely proportional to the bed height) the pressure drop is proportional to the bed height. Consequently the pressure drop through a bed of the small 15–5 pellets is the same as that which would occur across the deeper bed of 7·5 × 7·5 mm pellets which would be needed to achieve an equal performance in the same converter. Alternative catalysts range from 6 × 6 mm to 9 × 9 mm sizes.

At pressures around 20 atm and higher the initial pressure drop through the catalyst bed is small, and, on a well designed plant, is relatively inexpensive. The aim of the catalyst manufacturer should, therefore, be to allow the cost of the converter to be decreased. This can be done by decreasing the converter diameter and increasing its height—which raises the pressure drop but need not raise it to an unacceptable level. The high activity of catalyst 15–5 makes it well suited to small-sized converters, and its high strength ensures that the increase in pressure drop during long service is minimal. Future developments of high temperature shift converters will probably be towards higher pressures. The need is for reliability rather than increased conversion, because low temperature shift catalysts are better suited to high conversions. The operating temperature and the activity of present day catalytic material is adequate. Further adjustments in pellet size may be possible at higher pressures, allowing further decreases in converter sizes. The need for strong catalysts will continue, but there would be economic advantages in decreasing the bulk density if this could be achieved without lowering the strength.

Low-temperature shift (CO conversion) catalysts

With the advent of steam reforming over easily poisoned nickel catalysts, the production of an almost poison-free synthesis gas became economically attractive. This has resulted in an increased range of materials being catalytically suitable for synthesis gas production. In particular, it has become possible to develop the potential of copper. The arguments in Chapter 1 make copper an obvious choice for the CO conversion reaction because it is expected to have selectivity and activity at much lower temperatures than the conventional Fe_3O_4 catalysts. The literature records a long history of investigation into the catalytic properties of copper, but early workers always observed a rapid decline in activity due not only to its susceptibility to poisons but also to a rapid loss of surface area. Combinations of copper with zinc oxide have been used as organic hydrogenation-dehydrogenation catalysts for many years, and these catalysts, which have some of the properties of low-temperature shift-catalysts, formed a natural starting point for further research.

Formulation

The high activity of copper metal, described at the beginning of this chapter (p. 98) makes it a particularly suitable catalytic material for use at low temperatures where restrictions in conversion, due to equilibrium, are minimised. Pure copper metal, heated to temperatures at which reaction rates are adequate, suffers the disadvantage of losing surface area through thermal sintering, and 'spacer' materials must therefore be employed as described in Chapter 2. These materials are frequently termed supports, but a clear distinction must be made between inert carriers and spacers. Many materials have been tried, zinc oxide and alumina appearing to be the most effective, but chromia and other oxides also showing some success. The method of incorporating the support is important, since the object is to separate finely divided copper crystallites by particles of spacer material. Clearly this will not be achieved unless the particle size of the support relative to the copper crystallites is correctly chosen. An unsuitable formulation or choice of support results in inadequate protection for the copper surface, and much of the support merely dilutes the available quantity of active material. This is illustrated in fig. 54, where activity is plotted against time-on-line for several catalysts of varying composition and formulation. Clearly the catalyst in which zinc oxide and alumina are incorporated as spacers is the most satisfactory. The experiments made use of poison-free gases, and the observed effects are entirely the result of thermal sintering. This throws light on why catalysts of apparently similar composition can have widely different thermal stabilities. The potential life of the catalyst is, therefore, highly dependent on the spacer material having submicroscopic dimensions of a similar magnitude to those of the active catalytic

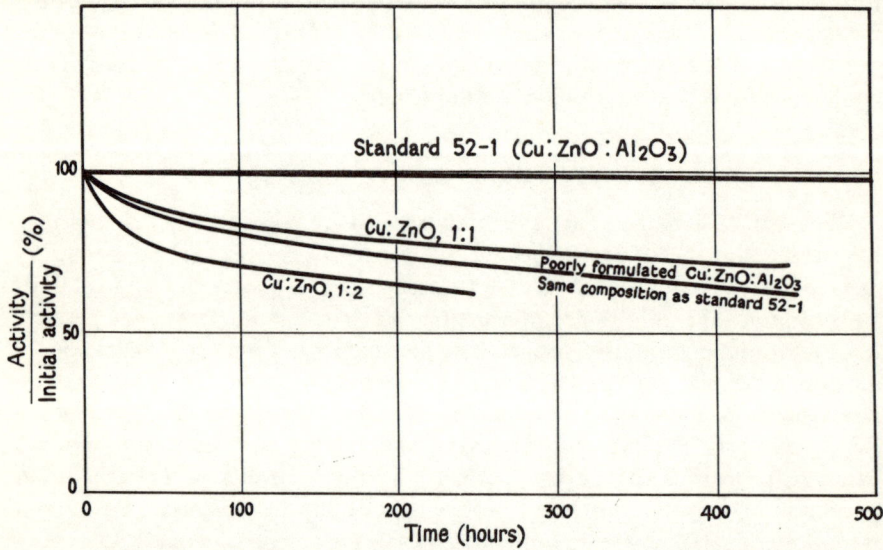

FIG. 54. Thermal stability of different catalyst formulations.

material, and the spacer must withstand operation under reaction conditions (Chapter 2, figs. 9 and 10).

The nature of the support and the method of formulation also affect the physical properties of the catalyst. Reference to Chapter 2, fig. 5, indicates how the special requirement of extra high activity can influence life and strength, making some compromise necessary. Catalyst 52–1 is designed for stability, because, in practical terms, a long lifetime is more useful than very high initial activity. Activity is related to high specific area and adequate pore volume. Fluid flow properties are related to strength (Chapter 2, fig. 5). Lifetime, being dependent on structural stability, is more related to the way in which the constituents are compounded than to variations in the list of ingredients. Catalyst 52–1 is composed of 30 per cent CuO, 45 per cent ZnO, and 13 per cent Al_2O_3. It has a surface area of 60 $m^2 g^{-1}$ and a pore volume of 0·4 $cm^3 g^{-1}$.

Reduction

Low-temperature shift catalysts, in their manufactured form, consist of copper oxide and either zinc oxide or alumina, preferably both. During reduction, with either hydrogen or carbon monoxide, the active phase is formed when the copper oxide is reduced to copper. It is easily shown from the appropriate equilibrium constants that zinc oxide (fig. 55) and alumina will not be reduced. The heat of reduction of the copper oxide corresponds to 21 kcal/mole of copper. Thus, because of the large amount of heat that can be liberated and the temperature sensitivity of copper catalysts, it is necessary to reduce them in a carefully controlled manner. Usually the catalyst temperature is kept below 250°C. This can be done by diluting the reducing gas (hydrogen) with an inert

Reaction	Temperature (°C)	K_p	$\Delta H_{25°C}$ kcal mole^{-1}
$CO + 3H_2 \rightarrow CH_4 + H_2O$	220	1.4×10^{10}	-49.3
$CO_2 + 4H_2 \rightarrow CH_4 + 2H_2O$	220	10^8	-39.5
$2CO \rightarrow C + CO_2$	220	5.4×10^4	-41.2
$CuO + CO \rightarrow Cu + CO_2$	220	4.3×10^{13}	-30.5
$CuO + H_2 \rightarrow Cu + H_2O$	220	3.1×10^{11}	-20.7
$ZnO + CO \rightarrow Zn + CO_2$	220	7.1×10^{-7}	15.5
$ZnO + H_2 \rightarrow Zn + H_2O$	220	5.2×10^{-9}	25.4
$Cu + HCl \rightarrow CuCl + \frac{1}{2}H_2$	220	6.5×10	-10.4
$Cu + 2HCl \rightarrow CuCl_2 + H_2$	220	10^{-2}	-5.1
$Cu + H_2S \rightarrow CuS + H_2$	220	7.8	-6.8
$2Cu + H_2S \rightarrow Cu_2S + H_2$	220	1.2×10^5	-14.2
$ZnO + 2HCl \rightarrow ZnCl_2 + H_2O$	220	1.8×10^7	-29.9
$ZnO + H_2S \rightarrow ZnS + H_2O$	220	7.9×10^7	-18.3
$ZnO + CO_2 \rightarrow ZnCO_3$	150	4.2×10^{-1}	-17
	175	1.4×10^{-1}	-17
	200	5.1×10^{-2}	-17
	225	2.1×10^{-2}	-17

FIG. 55. Thermodynamic equilibria for reactions over Cu:ZnO catalysts.

gas such as nitrogen. Initially a mixture with up to 0.5 per cent hydrogen in nitrogen is employed, and the concentration of hydrogen in the gas should be gradually increased as the reduction proceeds. In this way the rate of liberation of heat is controlled by the rate of addition of hydrogen.

Deactivation and poisoning

Low temperature shift catalysts gradually lose activity. At first this was thought to be caused by poisoning, sulphur being the main culprit, but closer examination of catalysts in operation revealed two further factors which play an important role in deactivation. These are thermal sintering and halogen deactivation. Sulphur poisoning can be completely eliminated by the use of adequate guards, and halogens can generally be avoided when the need for this is recognised.

The susceptibility to thermal sintering, however, is a basic property of the catalytic material. It is apparent from fig. 54 that with well-formulated catalysts thermal sintering can be neglected, provided a temperature of about 250°C is not exceeded.

The thermal stability of this type of catalyst can be markedly affected by the method of preparation (ref. 62), and catalysts which have exactly the same chemical composition can have very different thermal stabilities. In fig. 54 the activities of two such catalysts are plotted against the time-on-line and, although both catalysts were tested under identical operating conditions with poison-free gases, clearly one of the catalysts lost activity more rapidly than the other. It was shown, using gas chemisorption techniques and X-ray diffraction, that the behaviour of the inferior formulation is caused by rapid sintering of the active

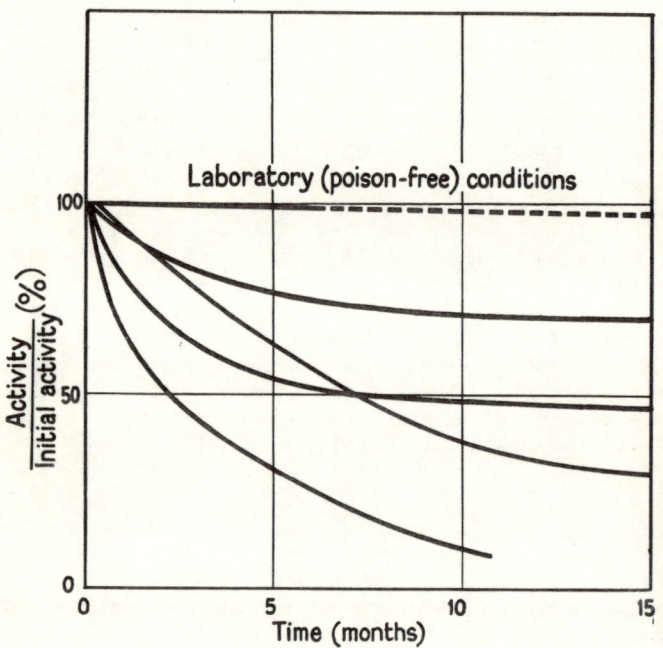

FIG. 56. Variation of activity with time under plant operating conditions.

species. Loss of activity caused by thermal sintering was found to be relatively unimportant with a well-formulated catalyst, certainly for operating temperatures up to 250°C. The difference in formulation was that the good catalyst contained 12 per cent spacer alumina. The other contained only 4 per cent spacer alumina together with 8 per cent finely ground alumina.

Fig. 56 shows the rate of loss of activity with time-on-line for identical catalysts operating under commercial conditions in various full-scale plants. The performance of an equivalent catalyst in a laboratory-scale reactor with poison-free gases is also given for comparison purposes. It is clear that the rate of loss of activity of a correctly formulated catalyst in the laboratory-scale unit is negligible compared to that in the full-scale plant. Since all the results shown were obtained over approximately the same temperature range, poisoning rather than thermal sintering would appear to be the more important factor contributing to loss of activity in the full-scale plant. This is in line with the fact that, after two months on line, no poisons could be detected by chemical analysis in catalyst discharged from the laboratory-scale units.

The distribution of poisons in catalyst discharged for examination from a commercial plant after 12 months operation is shown in fig. 57. The results indicate that most of the poisons are found at the top of the bed. Generally the level of sulphur at the top of a catalyst bed is around one per cent after 12 months operation, but recently, a catalyst examined after only nine months operation, showed a level of 2·6 per cent at the top of the bed.

Position of catalyst in bed	Poison (% w/w) SO_3	Cl	Activity relative to that at bottom of bed (%)
Top	1·4	0·08	15
	0·7	0·04	15
	0·4	<0·01	72
	0·2	<0·01	75
Bottom	0·2	<0·01	100

FIG. 57. Distribution of poisons through a catalyst bed.

Chlorine appears to be a much more virulent poison than sulphur, because later tests aimed at differentiating between the effects of the two showed that most of the losses in activity shown in fig. 57 were due to chlorine. The mechanism of poisoning could be a subject for further research, because little is known of it at the present time. One assumption is that poisoning occurs by a blocking-off of active sites.

Since such small quantities of poison lead to substantial deactivation, the proportion of active surface in the catalyst relative to the total weight must be very small. The atoms on the catalyst surface form only a small proportion of the total, a proportion which depends on crystallite size, and for a typical low temperature shift catalyst it is generally of the order of 0·5 per cent. Since these catalysts generally contain around 30 per cent (w/w) copper, it can be concluded that 0·1 per cent w/w poison could be sufficient to deactivate the catalyst surface severely—an inference which is in rough agreement with the results obtained for chlorine. In plant operation the effect might not be so catastrophic as outlined above, since most poisons, by definition, will react irreversibly with the constituents of the catalyst, and will be removed by the catalyst at the front of the bed, thereby giving some protection to the remainder. This still does not detract from the need to use poison-free gases. The activity of the catalyst can also be protected by a guard bed of some suitable material (for example, zinc oxide) in those cases where sulphur poisoning is a problem. The most effective protection is obtained by the installation of additional catalyst.

Reactions on low-temperature shift catalysts

The normal constituents of the gases which pass over low temperature shift catalysts are CO, CO_2, H_2, N_2, A, CH_4, and H_2O, together with small traces of catalyst poisons such as sulphur and chlorine. During shut-down and discharge, air is sometimes introduced into the converter in controlled amounts.

A knowledge of possible reactions between the components of the catalyst and these gases can be gained from the appropriate thermodynamic data (fig. 55), and helps to explain the behaviour of the catalyst under plant conditions. Some of these reactions are discussed in the next section. The main reaction, the CO shift reaction (1), needs more detailed study, because the rate of this reaction controls the performance of the converter. In a following section the mechanism of the reaction and some kinetic equations are discussed.

Reactions with the catalyst

From a thermodynamic standpoint there are alternative reactions with shift process gas. The formation of methane (p. 101, reaction 4) is more easily achieved than the shift reaction (and under some conditions carbon monoxide could disproportionate into carbon and CO_2 [p. 101, reaction (5)]). A prime consideration in the choice of an agent as an industrial catalyst is its ability to catalyse selected reactions, and copper is chosen for low temperature shift duties because it *is* selective, and can catalyse the shift reaction without methane or carbon being formed.

The low temperature shift catalyst is activated by reduction, during which its copper oxide content is reduced to copper. The other catalyst components, such as zinc oxide and alumina, do not take part in this reaction. The reduction reaction is exothermic (fig. 55), and care must be taken to ensure that the rate of hydrogen addition is controlled. The reduction procedure is described in Chapter 8. Possible reactions between the components of the catalyst and the reaction gases are important. The formation of zinc carbonate must be considered because at 160°C, for example, CO_2 at a partial pressure of 3 atm will react with zinc oxide and weaken the catalyst. At higher temperatures the partial pressure of CO_2 can safely be higher. The action of water and CO_2 on zinc oxide is similar to thermal sintering, but takes place at lower temperatures (Chapter 1), and the spacer qualities of zinc oxide will diminish if reaction conditions approach possible compound formation.

The form of alumina used in the catalyst must be one which does not react with process gases, but which does have good spacer properties. The question of inertness presents no problems. In ICI catalyst 52–1, where both zinc oxide and alumina are submicroscopic spacers, the alumina not only hinders the thermal sintering of copper, but also hinders reaction-sintering of zinc oxide, and increases the ability of the catalyst to withstand the operating conditions. At the end of its life, a low-temperature CO shift catalyst must be discharged from the converter and may be exposed to air. The oxidation of copper with oxygen is an exothermic reaction (fig. 55), and the potential for generating heat, is even greater than that during catalyst reduction. Since the catalyst is easily damaged by over-heating, and since there are possible pyrophoric hazards, the re-oxidation must be controlled by one of the techniques described in Chapter 8.

Reaction mechanism and kinetics

Published data on the kinetics of the water gas shift reaction at low temperatures on copper catalysts are summarised in fig. 58. In all the equations the rate of reaction increases with increasing concentration of both carbon monoxide and steam. All the equations contain a term that increases the rate when the temperature is raised, and a term that decreases the rate to zero when equilibrium is approached. The remaining terms in equations (2) and (3) relate to the mechanism of the reaction, and to the method of correcting for the effect of pressure. Thus Moe publishes a graphical dependence between activity and pressure, whereas Campbell and Metcalfe have derived equation (3) in this particular form for the reaction when it is diffusion-limited, and consequently have included the normal pressure term for diffusion-limited reactions.

Kinetic equation	Catalyst composition	Reference
$r = \exp\left\{15 \cdot 92 - \dfrac{7050}{T^\circ R}\right\}\left((CO-x)(H_2O-x) - \dfrac{(CO_2+x)(H_2+x)}{K}\right)$	Cu:Zn:Cr	63
$r = KP_{H_2O}\left(\dfrac{P_{CO}}{AP_{H_2O}+P_{CO_2}}\right)^m \left(1 - \dfrac{P_{CO_2}P_{H_2}}{KP_{H_2O}P_{CO}}\right)$	Cu:Zn:Cr	64
$r = \dfrac{K_1 P_{CO} P_{H_2O}^{\frac{1}{2}}}{P^{\frac{1}{2}}(1+K_2 P_{CO}+K_3 P_{CO_2})}\left(1 - \dfrac{P_{CO_2}P_{H_2}}{K_p P_{H_2O} P_{CO}}\right)$	Cu:Zn:Al	65

FIG. 58. Published kinetics for the water-gas reaction using copper catalysts.

There are two ways by which a catalyst can promote a redox reaction. One is for reactants to be absorbed on the catalyst surface with one or more of the absorbed species reacting. The second mechanism is for the catalyst to be alternately oxidized and reduced. The two mechanisms are illustrated below as two series of successive reactions.

absorption of reactants
$$CO + Z \rightleftharpoons ZCO \quad (12a)$$
$$H_2O + Z \rightleftharpoons ZH_2O \quad (12b)$$

reaction
$$ZCO + ZH_2O \rightleftharpoons ZCO_2 + ZH_2 \quad (12c)$$

desorption of products
$$ZCO_2 \rightleftharpoons Z + CO_2 \quad (12d)$$
$$ZH_2 \rightleftharpoons Z + H_2 \quad (12e)$$

$$CO + MO \rightleftharpoons MOCO \quad (13a)$$
$$MOCO \rightleftharpoons MCO_2 \quad (13b)$$
$$MCO_2 \rightleftharpoons M + CO_2 \quad (13c)$$
$$H_2O + M \rightleftharpoons MOH_2 \quad (13d)$$
$$MOH_2 \rightleftharpoons MO + H_2 \quad (13e)$$

If it is assumed that one reaction step limits the reaction rate, and that all other reactions are at equilibrium, then a kinetic equation can be derived from the reaction mechanism. Each step can, in turn, be considered limiting, and the series of kinetic equations can be compared with reaction rates measured over a range of conditions. This normally leads to the identification of the limiting step.

The two reaction mechanisms shown above lead to very similar kinetic equations. Thus reaction (12a) parallels (13a). Similarly, (12b) and (13d), (12d) and (13c), and (12e) and (13e), are also parallel reactions. Reaction (12c), however, involves the reaction between two absorbed species, and its rate depends on the concentration of ZCO and ZH_2O. Reaction (13b), on the other hand—the reaction between one absorbed species and the catalyst—is not dependent on H_2O, except through the competition between H_2O and CO to occupy catalyst sites. The second mechanism is very similar to the one proposed by Shchibrya, Morozov and Temkin (ref. 64) to explain their reaction kinetics in the second equation in fig. 58. A parallel result can be obtained from the first mechanism if step (12a) is assumed to limit the reaction. The Campbell–Metcalfe equation (fig. 58), derived after careful consideration of plant data and semi-technical results, suggests that the reaction on the catalyst is a first order reaction which includes both CO and H_2O, a step resulting from absorption (but not from a redox system) obtained from the first mechanism. The appropriate condition for this to take place is when reaction (12c) is the limiting step, and it can be shown that

$$-G\frac{dP_{CO}}{dV} = \frac{kA^2 KP_{CO} KP_{H_2O}}{(1+K_a P_{CO}+K_b P_{H_2O}+K_d P_{CO_2}+K_e P_{H_2})^2}\left(1-\frac{P_{CO_2} P_{H_2}}{K_p P_{CO} P_{H_2O}}\right)$$

where P_{CO}, P_{H_2O}, P_{CO_2}, and P_{H_2} are the partial pressures of the reactants,
$-GdP_{CO}/dV$ is the rate of reaction,
A is a constant representing the total number of active centres per unit volume of catalyst,
k is the forward reaction rate of equation (12c), K_a, K_b, K_d, K_e are equilibrium constants for reactions (12a), (12b), (12d), and (12e),
K_p is the equilibrium constant for the overall reaction.

This equation is applicable only when there are no diffusional limitations. In practice, diffusion limits the reaction rate, and the equation must be modified to take account of diffusion. The modification depends on whether Knudsen diffusion or pore-diffusion predominate. The dimensions of the catalyst and the rate of reaction both indicate pore-diffusion. A further consequence of pore-diffusion is that the reaction rate becomes related to total pressure, P_T. A term D/P_T is therefore included and it can be shown that

$$-G\frac{dP_{CO}}{dV} = \frac{(kK_a K_b D)^{\frac{1}{2}} AP_{CO} P_{H_2O}^{\frac{1}{2}}}{P_T^{\frac{1}{2}}(1+K_a P_{CO}+K_b P_{H_2O}+K_d P_{CO_2}+K_e P_{H_2})}\left(1-\frac{P_{CO_2} P_{H_2}}{K_p P_{CO} P_{H_2O}}\right).$$

By combining some of the constants and neglecting small terms, this expression can be reduced to the Campbell–Metcalfe form.

Development of new low-temperature shift catalysts

The requirement that a low-temperature shift catalyst must be active at temperatures in the range 200–250°C has resulted in manufacturers placing the main emphasis on catalyst activity rather than on strength or lifetime. The activity of new catalyst is good, and bed volumes could be decreased to about 30 per cent of their present sizes if initial activity could be stabilised and poisoning eliminated. Higher catalyst strength, particularly the ability to withstand condensing steam, would be an advantage, as would a lifetime in excess of two years. If the tendency for plant operating-pressures to increase continues, the steam condensation temperature will rise to 200–250°C, and there will be a need for catalysts to operate at somewhat higher temperatures.

The competitive nature of the catalyst business will ensure that customers will look for value for money in the different agents offered. The user may go either for cheap catalysts with inferior properties, or for tailor-made catalysts possessing advantages according to the requirements of his particular plant. However, the economics of most modern ammonia plants are linked to the inerts level in the synthesis section, and a relatively small improvement in CO conversion can soon repay the cost of the catalyst.

Catalyst 52–1 represents the results of many years research aimed at developing a structure in which copper is retained in a finely divided but stable form, and it is a useful low temperature shift catalyst. Further development along this theoretical path may, however, result in still better catalysts.

Methanation

Prior to ammonia synthesis, carbon monoxide and carbon dioxide in the synthesis gas must be either removed from the system, or converted to inert species, since any oxygen or oxygen compounds entering the ammonia loop will poison the synthesis catalyst. Carbon oxides in the gas are removed by chemical or physical absorption techniques, or alternatively they can be reacted to form water, and then condensed from the system. Modern plants, based on steam reforming, follow the combination of high and low temperature water gas shift with the absorption of the carbon dioxide. Subsequently methanation removes residual carbon oxides. Methanation is a simple process involving a minimum of plant and a relatively inexpensive catalyst.

Thermodynamic restrictions on reaction

In designing a methanator, the principal consideration is the exit concentration of carbon oxides, and this is related to the economics of the ammonia loop. In general the economic operating exit level from a methanator is a total carbon oxide concentration of not more than 5 ppm, or occasionally 10 ppm, at the end of the life of the methanation catalyst. This concentration plays a deciding role in the life of the ammonia synthesis catalyst (p. 139). On some plants where centrifugal compression is employed, with a kickback line from the discharge of the circulator to the inlet to the HP stage, ammonium carbamate may form, if the make-up gas contains more than 10 ppm of carbon dioxide. This may cause stress corrosion in the compressor.

The reacting species in the gas are hydrogen, carbon monoxide, and carbon dioxide, which may participate in the following reactions:

$$CO + 3H_2 \rightleftharpoons CH_4 + H_2O \tag{14}$$

$$CO_2 + 4H_2 \rightleftharpoons CH_4 + 2H_2O \tag{15}$$

$$2CO + 2H_2 \rightleftharpoons CH_4 + CO_2 \tag{16}$$

$$CO + H_2O \rightleftharpoons CO_2 + H_2 \tag{17}$$

$$2CO \rightleftharpoons CO_2 + C \tag{18}$$

$$CO + H_2 \rightleftharpoons H_2O + C \tag{19}$$

$$CH_4 \rightleftharpoons 2H_2 + C \tag{20}$$

Reaction (16) may be considered to be a combination of reactions (14) and (17).

Methanation equilibria

The principal reactions which occur during methanation are reactions (14) and (15). Both are exothermic.

Equilibrium constants for these reactions may be defined by

$$K_{p,CO} = \frac{P_{CH_4} P_{H_2O}}{P_{CO} P_{H_2}^3} \quad K_{p,CO_2} = \frac{P_{CH_4} P_{H_2O}^2}{P_{CO_2} P_{H_2}^4}.$$

These equilibria have been calculated for the temperature range 200–600°C, from thermodynamic data (see appendix), and are given in fig. 59. These data can be used to evaluate the concentrations of CO and CO_2 which would be expected under equilibrium conditions.

At an exit temperature of 325°C the equilibrium constant for CO methanation is $2 \cdot 19 \times 10^6$ atm^{-2}, and for CO_2 methanation is $7 \cdot 95 \times 10^4$ atm^{-2}.

A typical methanator inlet gas consists of CO, 0·5 per cent; CO_2, 0·2 per cent; H_2O, 1·0 per cent; H_2, 73·3 per cent; CH_4, 1 per cent; N_2, 24 per cent.

Since conversion of the carbon oxides to methane is almost complete, the exit gas contains about 1·7 per cent methane, 2 per cent water, and 70 per cent hydrogen. At atmospheric pressure the equilibrium concentrations of carbon monoxide and carbon dioxide are $P_{CO} = 3 \cdot 96 \times 10^{-10}$ atm, and $P_{CO_2} = 2 \cdot 88 \times 10^{-10}$ atm. That is to say, the CO concentration is $3 \cdot 96 \times 10^{-4}$ ppm, and the CO_2 concentration is $2 \cdot 88 \times 10^{-4}$ ppm. At higher pressures the equilibrium concentrations decrease. Clearly the performance of a methanation catalyst is unlikely to be limited by approach to equilibrium. At normal exit concentrations, methanation is a first order reaction, and the fall in reactant concentrations is the factor which limits the rate of reaction.

The exothermic methanation reactions of both carbon monoxide and carbon dioxide have heats of reaction ($\Delta H_{25°C}$) of $-49\cdot27$ and $-39\cdot44$ kcal/mole of carbon oxide respectively, so the temperature-rise corresponding to a given conversion of carbon oxides can be calculated, provided that heat losses from the converter can be estimated. Normally operation is adiabatic, since the heat loss from a well lagged plant converter is negligible compared with the heat input. Another justifiable assumption is that specific heats are constant over the normal range of operating conditions. The temperature rise for a typical methanator gas composition is 74°C per 1 per cent of carbon monoxide converted, and 60°C per 1 per cent of carbon dioxide converted.

	Equilibria for $CO + 3H_2 \rightleftharpoons CH_4 + H_2O$ and $CO_2 + 4H_2 \rightleftharpoons CH_4 + 2H_2O$	
$T(°C)$	$\dfrac{P_{CH_4} P_{H_2O}}{P_{CO} P_{H_2}^3}$	$\dfrac{P_{CH_4} P_{H_2O}^2}{P_{CO_2} P_{H_2}^4}$
200	0.21547×10^{12}	0.94748×10^9
220	0.23473×10^{11}	0.15589×10^9
240	0.30353×10^{10}	0.29435×10^8
260	0.45626×10^9	0.62706×10^7
280	0.78369×10^8	0.14863×10^7
300	0.15161×10^8	0.38747×10^6
320	0.32635×10^7	0.11001×10^6
340	0.77350×10^6	0.33737×10^5
360	0.20004×10^6	0.11094×10^5
380	0.56011×10^5	0.38882×10^4
400	0.16862×10^5	0.14442×10^4
420	0.54247×10^4	0.56582×10^3
440	0.18550×10^4	0.23282×10^3
460	0.67099×10^3	0.10023×10^3
480	0.25564×10^3	0.44995×10^2
500	0.10219×10^3	0.20997×10^2
520	0.42710×10^2	0.10157×10^2
540	0.18605×10^2	0.50814×10^1
560	0.84234×10^1	0.26225×10^1
580	0.39532×10^1	0.13936×10^1
600	0.19186×10^1	0.76104×10^1

FIG. 59. Equilibria for CO and CO_2 methanation for differing temperatures.

Under methanation conditions the only likely reactions are the methanation of both oxides of carbon, and either the water gas shift reaction or the reverse shift reaction. The conditions at the methanator inlet could give rise to carbon formation by reactions (18) and (19), but removal of carbon by reaction (20) could be a faster reaction. At the exit, and in the bulk of the catalyst, carbon deposition is thermodynamically impossible. In practice there is no tendency for carbon formation from carbon monoxide. Some general data relevant to methanation operations are included in fig. 60.

Methanation catalysts

Methanation of small amounts of carbon oxides has been an industrially important gas purification process for only about 10 years. Previously the methanation reaction had been applied to fuel gas production with a molar ratio of hydrogen to carbon monoxide of between 1:1 and 3:1.

Specific heats

(a) *Catalyst*

NiO $\quad C_p = 11\cdot3 + 0\cdot00215\ T$ cal/deg. C g mole
Ni $\quad\ \ C_p = 4\cdot26 + 0\cdot00640\ T$ cal/deg. C g mole
$Al_2O_3 \quad C_p = 22\cdot08 + 0\cdot008971\ T - 522500\ T^{-2}$ mole

At 300°C specific heats per g are:
NiO $\qquad$ 0·1678 cal/g °C
Ni $\qquad\ $ 0·139 cal/g °C
Al_2O_3 $\quad\ $ 0·251 cal/g °C

For unreduced catalyst containing 30 per cent NiO and 70 per cent alumina, at 300°C: $C_p = 0\cdot23$ cal/g °C.
For a similar catalyst in the reduced state: $C_p = 0\cdot22$ cal/g °C.

(b) *Gas stream*

$H_2 \quad\ \ C_p = 6\cdot62 + 0\cdot0081\ T$ cal/g mole °C
$N_2 \quad\ \ C_p = 6\cdot50 + 0\cdot00100\ T$ cal/g mole °C
CO $\quad C_p = 6\cdot6 + 0\cdot00120\quad$ cal/g mole °C
$CO_2 \quad C_p = 10\cdot34 + 0\cdot00274\ T - 195500\ T^{-2}$ cal/g mole °C
$CH_4 \quad C_p = 5\cdot34 + 0\cdot0115\ T$ cal/g mole °C
$H_2O \quad C_p = 8\cdot22 + 0\cdot00015\ T - 0\cdot00000134\ T^2$ cal/g mole °C

For synthesis gas at 300°C $C_p = 0\cdot316$ kcal/NM^3 °C $= 0\cdot01097$ Btu/Nft^3 °F

Specific heats of gas mixtures may be calculated assuming that the contribution of each component to the total specific heat is the product of the mole fraction of that component and its specific heat.

Heats of formation

NiO $\qquad \Delta H_f^{298} - 58\cdot4$ kcal/g mole
$H_2O \quad\ \ \Delta H_f^{298} - 57\cdot7979$ kcal/g mole
CO $\qquad \Delta H_f^{298} - 26\cdot416$ kcal/g mole
$CO_2 \qquad \Delta H_f^{298} - 94\cdot052$ kcal/g mole
$CH_4 \qquad \Delta H_f^{298} - 17\cdot889$ kcal/g mole

Heats of reaction

	Heat of reaction at	
	298°C	573°K
$CO + 3H_2 \rightarrow CH_4 + H_2O$	−49·27	−51·83
$CO_2 + 4H_2 \rightarrow CH_4 + 2H_2O$	−39·43	−41·92
$CO + H_2O \rightarrow CO_2 + H_2$	− 9·85	− 9·91
$NiO + H_2 \rightarrow Ni + H_2O$	+ 0·61	− 0·84
$Ni + \frac{1}{2}O_2 \rightarrow NiO$	−58·4	−58·3

FIG. 60. Thermodynamic data relevant to methanation.

Catalyst formulation

Early work on methanation was largely restricted to carbon monoxide, but it was found that catalysts which were active for this reaction also catalysed the hydrogenation of carbon dioxide. The very early experimental work mainly involved the use of nickel catalysts, although some work was done on the other group VIII metals. Subsequently, iron catalysts were extensively studied, but were found to be subject to excessive carbon deposition, leading to blockage of the catalyst pores and consequent deactivation. In addition, iron catalysts showed a tendency to form higher hydrocarbons which appeared as a liquid product.

Precious metals, particularly ruthenium, have considerable methanation activity, and catalysts are available containing about 0·5 per cent of the metal supported on alumina. These catalysts can operate at low temperatures, but under normal conditions are no more active than conventional nickel catalysts, and are generally too expensive to be used in large ammonia plants.

Nickel catalysts have proved more active than iron for methanation of carbon oxides, and they are much more selective, eliminating the problem of carbon deposition and hydrocarbon formation. Most commercial methanation catalysts contain nickel as the active phase, supported on an inert substrate such as alumina, kaolin, or calcium aluminate cement. Some formulations contain either magnesia or chromia as promoters or stabilisers.

The nickel oxide content of the catalyst is a factor in determining the activity of the catalyst, but the reducibility of the nickel oxide is equally important. In a poorly prepared catalyst only a proportion of the nickel oxide can be reduced by the normal reduction procedure, and the activity is relatively low. For example a nickel oxide-alumina spinel, or its precursor, is not completely reduced to nickel below temperatures of 400–500°C. Other oxides, such as magnesia, can react with nickel oxide to form solid solutions which are difficult to reduce. These factors, together with the physical properties of the material, affect the choice of catalyst formulations.

Kinetics of methanation

The methanation of carbon monoxide has been widely studied, and is well documented, while the methanation of carbon dioxide (particularly in the presence of carbon monoxide) has been largely neglected.

The rate of methanation of carbon monoxide at sufficiently low concentrations is found to be first order with respect to carbon monoxide, and similarly, in the absence of carbon monoxide, the rate of methanation of carbon dioxide is first order with respect to carbon dioxide. When both carbon monoxide and carbon dioxide are present in the gas, carbon monoxide methanation is independent of carbon dioxide concentration. There is, however, an interaction which arrests the methanation of carbon dioxide until the carbon monoxide has been reduced to about 200–300 ppm. For this reason, in mixtures containing both carbon oxides, carbon dioxide is more difficult to methanate than carbon monoxide.

Methanation catalysts have very high intrinsic activities under non-diffusion-limited conditions. Consequently the pelleted catalyst, under plant conditions, is very strongly diffusion-limited. The activation energies for methanation of carbon monoxide and carbon dioxide over nickel catalysts are similar, the published value for carbon dioxide methanation being 7 kcal g. mole^{-1}, and the observed value for carbon monoxide being 7·4 kcal g. mole^{-1}.

In addition to the direct methanation of carbon monoxide and carbon dioxide, there is evidence that carbon dioxide is removed by the reverse water gas shift reaction, by conversion to carbon monoxide:

$$CO_2 + H_2 \rightleftharpoons CO + H_2O.$$

If a catalyst is tested with carbon dioxide, with no carbon monoxide in the inlet gas, a trace of carbon monoxide is found in the exit gas, indicating the occurrence of the above reaction.

The methanation reactions are first order with respect to carbon oxides, but do not show a first-order dependence on total pressure, because of diffusional and retardation effects. The overall pressure dependence of the rates is roughly proportional to P to a power between 0·2 and 0·5, and for a given catalyst it appears to be the same for both methanation reactions.

Catalyst reduction

Methanation catalysts are manufactured as supported nickel oxides, and must be reduced to nickel to make them active. The usual method of reduction is by process gas. A complete reduction schedule is described in Chapter 9. The reduction of nickel oxide can occur by two reactions:

$$NiO + H_2 \rightarrow Ni + H_2O \qquad \Delta H_{25°C} = +0\cdot61 \text{ kcal}$$

$$NiO + CO \rightarrow Ni + CO_2 \qquad \Delta H_{25°C} = -7\cdot23 \text{ kcal}$$

Neither of these reactions is strongly exothermic, and the reduction process itself does not cause a large temperature rise in the catalyst bed. Once some metallic nickel has been formed by reduction with process gas, however, methanation will start, giving the corresponding temperature rise. For this reason the gas used for reduction must contain as little CO and CO_2 as possible—certainly not more than 1 per cent in total. It is worth making checks to ensure that the carbon oxides concentration does not increase during the reduction, due, for example, to malfunction of the carbon dioxide removal unit, because these precautions protect both the catalyst and the converter. Catalyst deactivation occurs at 500–600°C depending upon the catalyst formulation. The maximum design temperature for converters is frequently about 450°C.

In the later stages of reduction, it is usually advantageous (though not essential with catalyst 11–3) to raise the temperature to about 400°C because this increases the proportion of reduced nickel (see last section). To achieve this temperature it may be necessary to increase the carbon monoxide content of the inlet gas by a controlled by-passing of the low temperature shift catalyst. This technique is often the only means for providing extra heat to the catalyst when the gas entering the methanator is heated by exchange with the exit gas.

Catalyst performance

Catalyst volumes are designed to give a specified performance at the end of the life of the catalyst, and as a result new catalyst generally markedly excels the prescribed performance. A modern ammonia plant of typical design performs as follows (at the end of the catalyst's useful life):

Inlet gas composition		
	CO	0·5%
	CO_2	0·2%
	CH_4	1·0%
	H_2	73·3%
	N_2	25%

Exit gas composition	$CO + CO_2$	5 ppm
Inlet temperature	315°C	
Exit temperature	364°C	
Space velocity	4000–4500 h^{-1}	
Pressure	30 atm	

The actual values for space velocity and inlet temperature are interrelated and depend on the overall heat recovery of the plant. There are savings in catalyst usage and in converter size when the inlet temperature is raised. An inlet temperature of 350°C gives the lowest capital cost, but lower inlet temperatures sometimes permit savings in running costs and give the plant more flexibility with respect to increased quantities of CO and CO_2 in the inlet gas.

Fresh catalyst usually reduces the carbon oxides to considerably less than 5 ppm, and the inlet temperature to the bed can be lowered to 270–280°C without exceeding the design concentration. Operation at the lowest convenient temperature sometimes economises on heat requirements, and also permits operation with higher carbon monoxide and carbon dioxide concentrations in the inlet gas. If the carbon monoxide concentration in the inlet gas is increased slightly, the exit temperature increases correspondingly, the rate of reaction increases, and the exit carbon oxides concentration normally falls. A similar effect is observed when the inlet carbon dioxide is increased slightly (provided that the inlet carbon monoxide is not too high), although the temperature rise in this case is lower. Because of the interaction between carbon monoxide and carbon dioxide, large increases in either the carbon monoxide or carbon dioxide inlet concentrations both lead to increases in the exit carbon dioxide concentration.

The conditions under which a methanator is run depend not only on the activity of the methanation catalyst, but also on the activity of the low-temperature shift catalyst, and the efficiency of the carbon dioxide removal unit. With a new low-temperature shift catalyst the carbon monoxide entering the methanator may represent only 0·2–0·3 per cent of the gas, and the temperature rise in the methanator will be correspondingly low. To obtain satisfactory conversion in the methanator under these conditions it is often necessary to operate at a temperature 10–20°C higher than that used when the inlet carbon monoxide concentration is 0·5 per cent.

When the gas entering the methanator is heated by that leaving, a high activity low temperature shift catalyst sometimes lowers the carbon monoxide concentration, and hence the temperature rise in the methanator, to such a degree that a satisfactory exit temperature cannot be achieved. To restore operation the carbon monoxide concentration in the gas entering the methanator can be increased by allowing some gas to by-pass the low-temperature shift converter. This restores the performance of the methanator, but only at the expense of increasing the inerts content in the synthesis gas.

In the event of a temperature runaway in the methanation catalyst, caused by abnormally high carbon monoxide or carbon dioxide concentrations in the inlet gas, it is necessary to protect both the vessel and the catalyst from serious damage. The vessel should immediately be isolated on the inlet side, and then blown down to atmospheric pressure as quickly as possible. This has two advantages, it lowers the quantity of gas available for reaction, and at low pressure high temperatures are less dangerous to the vessel. If a nitrogen supply is available, the vessel should be purged with as large a flow as possible to accelerate cooling, but air must be excluded from the vessel because the exothermal nature of the oxidation reaction in the catalyst would cause a further rise in temperature. Methanation catalysts are not seriously damaged by steam or water, and therefore steam could be used as a purge instead of nitrogen, although its cooling effect is much less because of its relatively high temperature. Occasionally, rapid cooling is required when the catalyst temperature is below 100°C, and then water can be used, provided it is free from sulphur and chlorine compounds. Water is never used when the catalyst temperature is above 100°C because there is a danger of excessive pressure being developed.

Catalyst die-off

The performance of a methanation catalyst can normally be expected to be maintained for two to three years. Under favourable conditions the catalyst may give even longer service. The two principal causes of loss of activity are poisoning and sintering. The only other potential cause of catalyst failure is breakdown of the pellets, resulting in an excessive pressure drop.

The poisons most likely to be encountered on an ammonia plant are compounds originating in the carbon dioxide removal plant. It is almost inevitable that there will be a small amount of carry-over of liquid into the methanator, but in general this does not have serious consequences. Plant malfunction can be much more serious, and large quantities of carbon dioxide removal liquor are sometimes pumped over the catalyst. Fig. 61 shows the effect of the commoner CO_2 removal media on methanation catalyst activity.

Because the low temperature shift catalyst is itself poisoned by sulphur, and contains zinc oxide, which is a good sulphur absorbent, it is unlikely that the methanation catalyst, in normal operation, will be exposed to sulphur in the

Benfield process	Aqueous potassium carbonate	Blocking of pores of methanation catalyst by evaporation of potassium carbonate solution.
Vetrocoke process	Aqueous potassium carbonate–arsenious oxide	As Benfield. Also As_2O_3 is poison—about one-half of activity is lost when As = 0·5 per cent
Benfield DEA	Aqueous potassium carbonate plus 3 per cent diethanolamine	As Benfield. DEA is harmless.
Sulphinol	Sulpholane, water, di-isopropanolamine	Sulpholane will decompose and give sulphur poisoning.
MEA, DEA	Mono- or diethanolamine in aqueous solution	No poisoning effect.
Cold Rectisol	Methanol	No poisoning effect.

FIG. 61. Poisoning effects on CO_2 removal systems.

gas stream. If, however, the low temperature shift converter is partially bypassed, there is a possibility of sulphur reaching the methanation catalyst which may become poisoned. Serious deactivation of catalysts containing about 30 per cent nickel oxide (before reduction) occurs when the sulphur content exceeds about 0·1–0·2 per cent.

There is also a possibility that sulphur may reach the methanation catalyst via the carbon dioxide removal liquor, if gas is passed through the carbon dioxide removal unit during the start-up period of either reforming or high temperature shift catalysts. These catalysts evolve sulphur during and immediately after reduction, as discussed under 'Reactions of sulphur compounds', above (see pp. 102–104). And if the gas from them is passed through carbon dioxide removal liquor, the sulphur will be absorbed in the liquor. Subsequently this sulphur will be desorbed into the gas, and may poison the methanation catalyst. Halides poison nickel catalysts, but are not normally encountered in methanation catalysts, and would be introduced only if water containing chloride was passed over the catalyst.

As the poison level in the catalyst builds up, the apparent activity of the catalyst declines, partly due to the reduction in the unpoisoned nickel surface, and partly due to a reduction in activation energy caused by pore-mouth poisoning. If the inlet temperature remains constant, then the exit carbon oxides concentration will rise, or alternatively the inlet temperature may be increased to maintain a constant exit concentration.

At temperatures below 400°C the rate of die-off is low, and modern methanation catalysts can withstand high temperatures better than the methanation vessel. Consequently overheating is not a common cause of loss of activity.

CHAPTER 7

Ammonia synthesis catalysts

THE synthesis of ammonia from nitrogen and hydrogen at pressure is a 'classic' in the field of applied chemistry. The process was introduced in about 1910 by BASF in Germany, and was one of the first large-scale applications of catalysis, following a few years after SO_2 oxidation and ammonia oxidation. The successful introduction of the process on an industrial scale demonstrated for the first time the value of the application of thermodynamic and kinetic principles to chemical reactions.

The catalyst is also a classic example of a heterogeneous catalyst, with one major component more effective than any other element, but requiring the addition of other compounds as promoters and stabilisers to increase its effectiveness over long periods of time. Over the 50 years or more since the process was introduced the catalyst has been the subject of continuous study in both academic and industrial laboratories. These studies have thrown a lot of light on its mode of action, and on the function of the promoters, and have in fact provided ideas that have been applied in many other fields of catalysis.

Thermodynamics of the process

The synthesis reaction is exothermic and is accompanied by a decrease in volume at constant pressure

$$N_2 + 3H_2 \rightleftharpoons 2NH_3 \qquad \Delta H_{500} = -26 \text{ kcal.}$$

The formation of high concentrations of ammonia, therefore, is favoured by operation at high pressure and low temperature. In practice over the last 50 years the optimum pressure for economic operation has been in the range 150–350 atm. Higher pressure processes up to 1000 atm are in operation, but normally the equilibrium advantages of very high pressure are more than offset by the higher cost of gas compression and by higher capital costs of plant.

The temperature at which the process is operated is determined by the characteristics of the catalyst. Thermodynamically, a low temperature is advantageous, but for kinetic reasons (that is, in order to obtain a reasonable rate of reaction) a rather higher temperature may have to be used. The most effective catalyst is clearly the one which will give a reasonable rate of conversion to ammonia at the lowest temperature. These considerations apply throughout the catalyst bed. As the synthesis reaction proceeds the temperature rises, making the reaction faster, but the equilibrium becomes less favourable, and the synthesised ammonia increases the rate of the reverse reaction. Careful control of the temperature profile through the bed is necessary if the optimum balance is to be obtained between the limits set by thermodynamic equilibrium and by the kinetics of the catalysed reactions in both the forward (synthesis) and reverse (ammonia decomposition) directions. As the ammonia concentration increases through the catalyst bed, the temperature for the maximum rate of conversion to ammonia becomes progressively lower. In practice, therefore, provision is made for the temperature to be decreased.

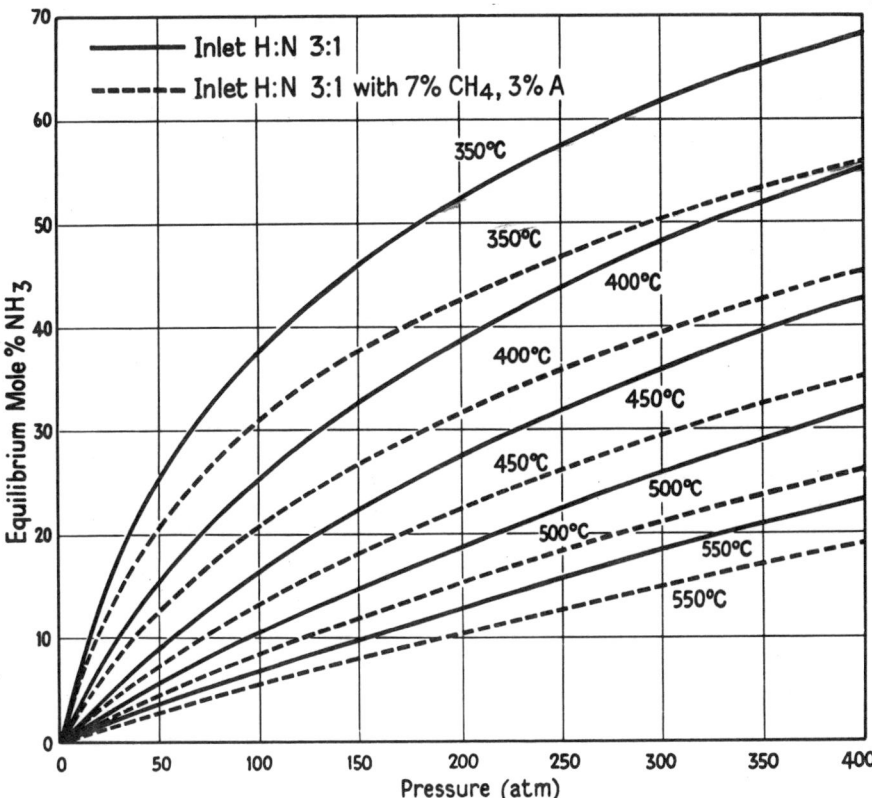

FIG. 62. Effect of pressure, temperature and inert gas on equilibrium ammonia concentration.

The thermodynamics of the process are not summarised in values of the equilibrium constant K_p, because, at the high operating pressures, the gases are non-ideal, and K_p is not only a function of temperature and pressure, but also depends on the $H_2:N_2$ ratio. Useful data have been determined experimentally by Haber (ref. 66) and by Larson and Dodge (ref. 67). These data have been analysed by Gillespie and Beattie (ref. 68) who developed a method for calculating the equilibrium composition of H_2, N_2, NH_3 and inert gas mixtures. Fig. 62 shows the equilibrium ammonia percentage as a function of temperature and pressure for 3:1 $H_2:N_2$ mixtures containing initially 0 per cent and 10 per cent inert gas.

Calculations on the converter system require a knowledge of the heat of reaction, and of the specific heats of the various gaseous mixtures. Because of non-ideality at the temperatures and pressures used commercially, the specific heats and the heat of reaction are a function of pressure as well as temperature. In a strict analysis, account should be taken of the heat of mixing of ammonia with unconverted synthesis gas. Figs. 63–66 show specific heat data for H_2, N_2, NH_3, and CH_4 as functions of temperature and pressure. A summary of values of heats of reaction at 500°C from several authors is given by Nielsen (ref. 69), the values depending upon the correction for heat of mixing. In practice it is convenient to work with a standard heat of reaction at 450°C (a reasonable

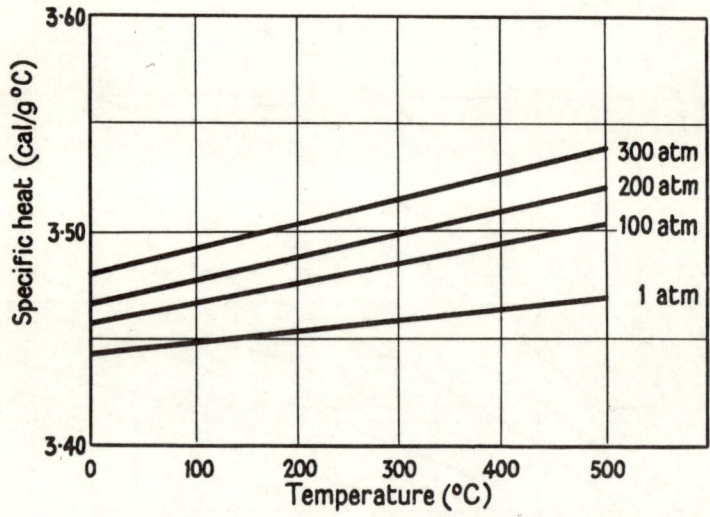

FIG. 63. Specific heat of hydrogen.

average temperature in commercial synthesis), ΔH_{450}, and an average specific heat $\bar{C}_p$, of the gas mixture. ΔH_{450} is taken as 12·95 kcal/mol NH_3 and $\bar{C}_p$ is calculated from

$$\bar{C}_p = \sum_i C_p(p_i) \frac{p_i}{P}$$

where $C_p(p_i)$ is the specific heat of component i at pressure p_i, its partial pressure in the gas mix. Thus in calculating the temperature rise (ΔT in °C) as a function of ammonia mol fraction, Z, in adiabatic synthesis from Z_i to Z_e we have

$$\Delta T = \frac{\Delta H_{450}}{\bar{C}_p} \left\{ Z_e \frac{(1+Z_i)}{(1+Z_e)} - Z_i \right\}.$$

With ΔH at 12·95 kcal/mol NH_3, this formula gives very good agreement with adiabatic temperature rises observed in both full scale and pilot plants.

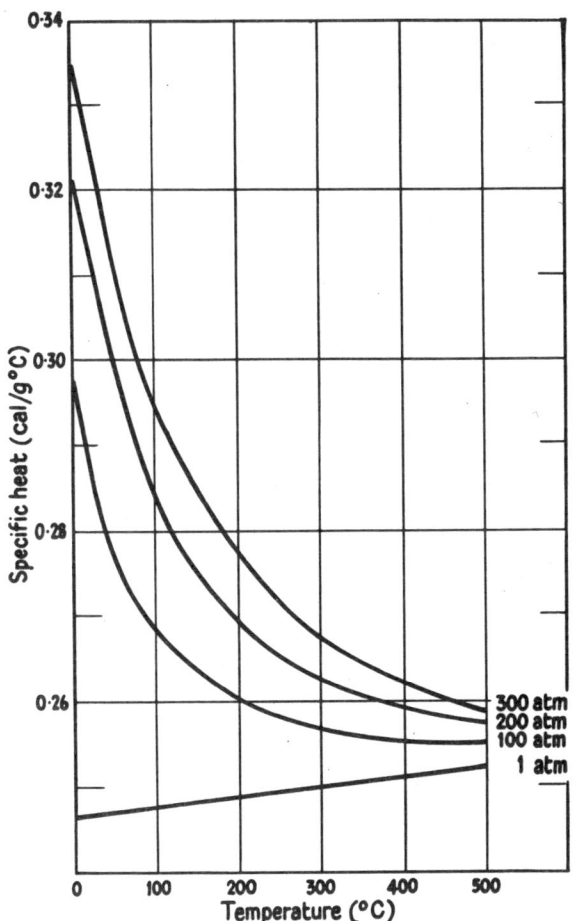

FIG. 64. Specific heat of nitrogen.

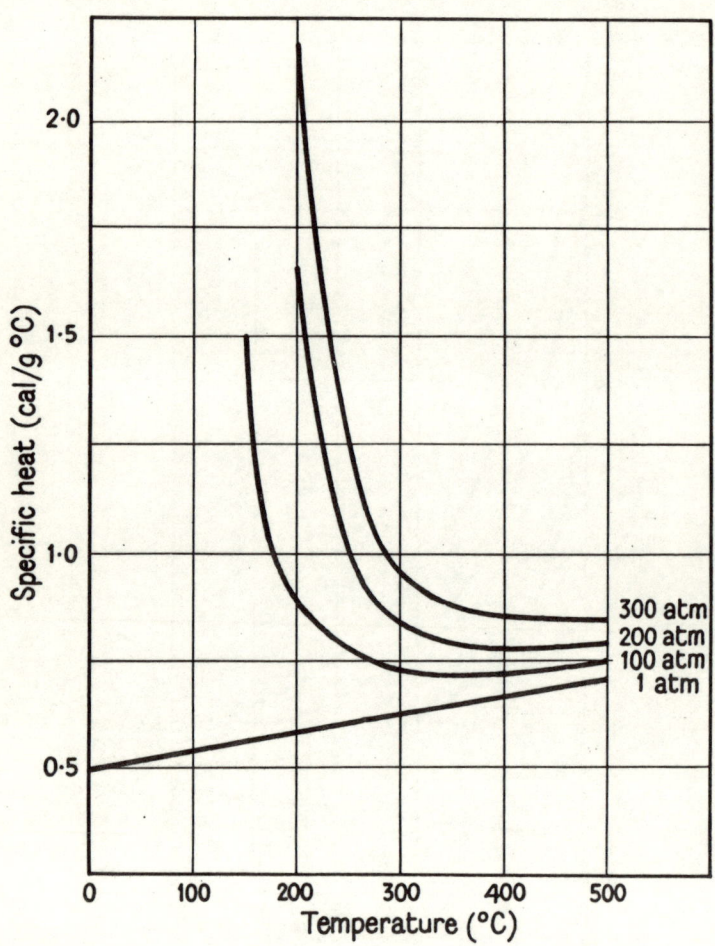

FIG. 65. Specific heat of ammonia.

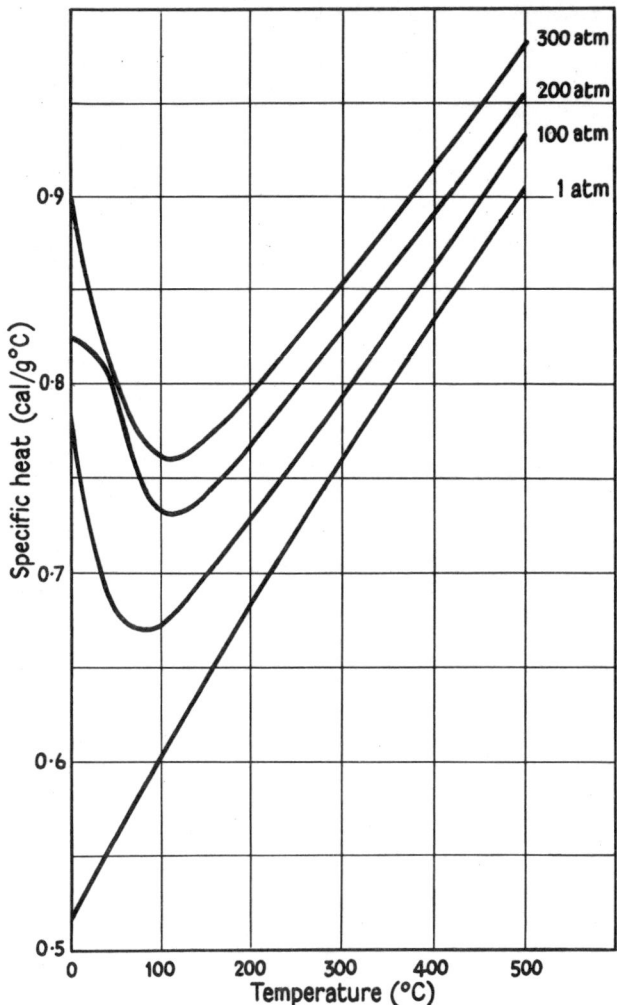

FIG. 66. Specific heat of methane.

Catalyst formulation

Synthesis catalysts throughout the world are now based on metallic iron promoted with alkali metals such as potassium, and various other oxides such as alumina or magnesia. The principle material used to make most commercially available catalysts is usually magnetite, Fe_3O_4, and some of the components in the catalyst originated as impurities in the magnetite. ICI catalyst 35–4 contains approximately 0·8 per cent K_2O, 2·0 per cent CaO, 0·3 per cent MgO, 2·5 per cent Al_2O_3, 0·4 per cent SiO_2 as well as traces of TiO_2, ZrO_2, and V_2O_5. In the making of catalysts of this sort it has to be recognised that these minor components have a big effect on the performance of the final catalyst, some beneficial, some harmful. They also interact themselves and hinder or assist each other. In a properly designed catalyst such as catalyst 35–4, these factors have been taken into account to optimise performance with a combination of high activity and long life.

Iron, the principal component

It is remarkable that the main component of the catalyst, iron, has remained unchanged since the catalyst was first introduced in 1914 by BASF in Germany. Since that date a considerable amount of research has been done on the catalyst, all of which has confirmed that iron is the best metal for the purpose, and certainly the cheapest. In the earliest researches by Haber and Mittasch other metals such as osmium and uranium were found to be as effective as iron, but they are more costly and introduce health hazards. Pure iron is a very effective catalyst, but it quickly loses its activity unless, as Mittasch found, other promoter oxides are present. It was found that the level of activity of the iron catalyst was increased by the addition of potassium, and these early researches also revealed the deleterious effects of gaseous poisons, such as oxygen and sulphur compounds, on the activity of the catalyst, which could be avoided by using rigorously purified gases.

The early discoveries were essentially empirical, but are still very relevant to modern synthesis catalyst. Research since Mittasch's time has been concerned with the elucidation of these discoveries, and in explaining the action of promoters, of poisons, and so on, and it is in this context that the theoretical considerations of Chapter 1 have proved valuable. A fairly comprehensive picture of the catalyst has been put together as a result of X-ray and microscopy studies, from which the crystalline species can be identified, and from gas absorption studies which throw some light on the distribution of these components.

As we have seen, thermodynamics indicates that the synthesis reaction is favoured by low temperature and high pressure, but the reaction does not take place in the absence of a catalyst. This is because of the high stability of the nitrogen molecule, associated with the high energy required to break the nitrogen-nitrogen bond. The catalyst functions by forming a nitrogen compound on the catalyst surface which is hydrogenated to ammonia. The metal-nitrogen bond is sufficiently weak, however, to enable the synthesised ammonia to desorb. The bond is too strong in elements such as lithium, calcium and aluminium, which form bulk nitrides direct from nitrogen. In the first series of transition metals, the optimum between surface nitride formation and surface ammonia desorption is obtained with iron, which does not form a nitride directly from nitrogen except at very high pressures (ten times higher than synthesis pressures), but forms it readily by reaction with ammonia. Nevertheless, iron chemisorbs nitrogen rapidly, and it is this adsorption which is generally considered to be the step which controls the rate of the overall synthesis process. In the higher transition series, ruthenium and osmium do not form bulk nitrides, and are effective synthesis catalysts.

After many years of research, iron is still the cheapest and most effective catalyst. Almost all of the synthesis catalysts are made by fusing magnetite, to which the required amounts of promoters have been added, and then cooling the molten mix by pouring it out as a shallow layer which, after solidification, is broken up and screened into the required size. Before use the catalyst is reduced to metallic iron by hydrogen or synthesis gas, either in the plant converter or by a pre-reduction and stabilisation process before charging.

FIG. 67. Magnetite crystals in unreduced synthesis catalyst. ($\times 10$.)

Magnetite, Fe_3O_4, has a spinel structure, similar to $MgAl_2O_4$, consisting of cubic packing of oxygen ions, in the interstices of which Fe^{2+} and Fe^{3+} ions are distributed. In the unreduced catalyst many of the crystals are large (see fig. 67). During reduction all the oxygen is removed, but no shrinkage occurs, so that very porous iron is obtained, occupying the same total volume as the original magnetite (see Chapter 2, fig. 13). This porosity is an important factor affecting the activity of the final catalyst. Another major factor is the size of the individual crystals of iron produced by the reduction processes, and this is largely determined by the nature and amounts of promoters which are present.

The promoters

In the manufacture of the catalyst, while the mix is fused, some of the alumina and magnesia promoters dissolve in the magnetite, Al^{3+} and Mg^{2+} replacing Fe^{3+} and Fe^{2+} respectively in the magnetite lattice, as they have similar

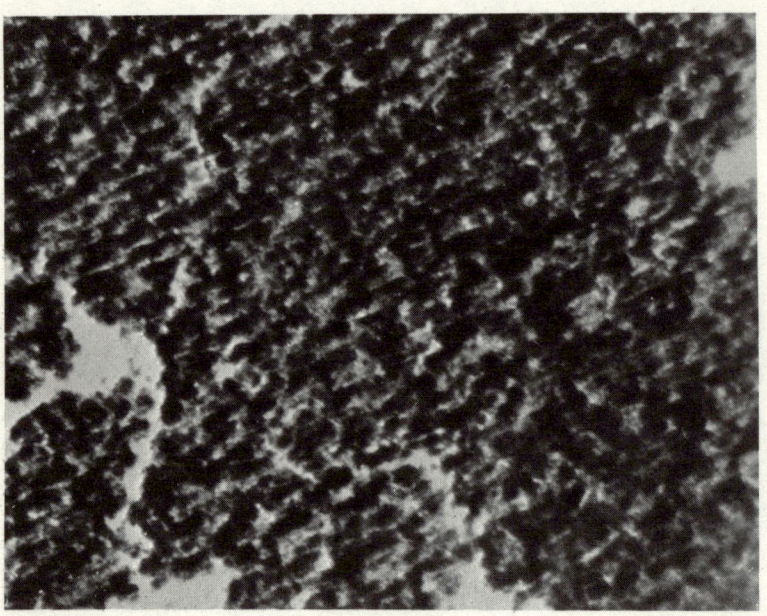

FIG. 68. Iron crystallites in a thin section of reduced catalyst. Electron micrograph of extraction replica. ($\times 50\,000$.)

ionic dimensions. During reduction, which starts on the outside of the Fe_3O_4 granules, the dissolved Al_2O_3 and MgO come out of solution and separate in the pores between the iron crystallites, hindering their further growth during the reduction and in subsequent use. This regular array of iron crystallites is shown in fig. 68. The crystallite size is very small, being between 200 Å and 400 Å for catalyst 35–4, the pores being of similar size. The corresponding surface area of the reduced catalyst can be as high as 15–20 m^2/g compared with less than 1·0 m^2/g for the unreduced catalyst. Promoters such as Al_2O_3 therefore assist in both the formation and in the preservation of small iron crystals of high surface area. Pure, unpromoted iron has a surface area of less than 1 m^2/g.

A high surface area is conducive to high activity (see p. 20), although, because of the complexity of the promoter chemistry, there is not necessarily direct proportionality between activity and total surface area. The synthesis reaction depends on gaseous diffusion within the pores (Chapter 2, fig. 7), as is discussed later, and the space between iron crystallites in the fine iron structure can easily be blocked by promoters. The promoters also cover a big proportion, up to 90 per cent, of the iron surface, and excessive amounts of promoters, as would be expected, have a deleterious effect on the activity. The maximum iron surface area is obtained in catalyst 35–4, which contains about 2·5 per cent of Al_2O_3.

Potash plays a different role—it increases the intrinsic activity of the iron surface, but also tends to decrease the iron surface area, so that again there is an optimum concentration (around 0·8 per cent K_2O for ICI catalyst 35–4) at which maximum activity is obtained. Potash does not dissolve in the magnetite during the fusion process. Some of it reacts with excess alumina and silica to form glassy phases of alumino-silicates between the magnetite crystals (fig. 69). These phases are stable on reduction, and the potassium they contain is ineffective in terms of promoter action. Some of the potash reacts with the magnetite to form potassium ferrites which are uniformly distributed in the phases between the Fe_3O_4 crystallites, and which on reduction produce potassium uniformly throughout the whole of the porous iron structure.

Calcium oxide, and other basic promoters (except MgO which dissolves in the magnetite) react in the first place with alumina and silica to form the glassy alumino-silicate type of compound, and some calcium ferrite. This leaves much of the K_2O available to activate the Fe. Possibly as a result of forming these compounds, CaO enhances the action of Al_2O_3 in stabilising the iron surface area and preventing sintering. It also makes the catalyst more resistant to poisoning by sulphur and chlorine. Some excess CaO dissolves in the magnetite, this tendency being most marked when the ferrous content of the magnetite is high and the ferric/ferrous iron ratio is below 2·0. In that case much of the potassium is rendered inactive by reaction with silica, and a catalyst of low activity is obtained. This may account for the commonly recognised fact that the most active catalysts are those produced from magnetite with a ferric/ferrous ratio of two or more.

Silica and other acidic components are commonly present as are impurities in the magnetite, and have the effect of 'neutralising' the K_2O, and other basic components such as CaO which lowers catalyst activity. Silica, however, has a stabilising effect, like Al_2O_3, so that high-silica catalysts tend to be more resistant to both water poisoning and sintering.

There is obviously considerable interaction between the various components of the catalyst, so that the optimum quantities have to be determined for each combination, taking account of impurities. The method of manufacture also has to be considered, because factors such as rates of cooling, and so on, affect the structure of the catalyst. Catalysts such as 35–4 are designed in this way, so their activities are associated with a high degree of stability and resistance to poisoning.

The distribution of the different crystalline phases in the unreduced and reduced catalysts can be determined by electron probe analysis (fig. 69). The amount of surface of iron and of promoters can be measured separately by gas

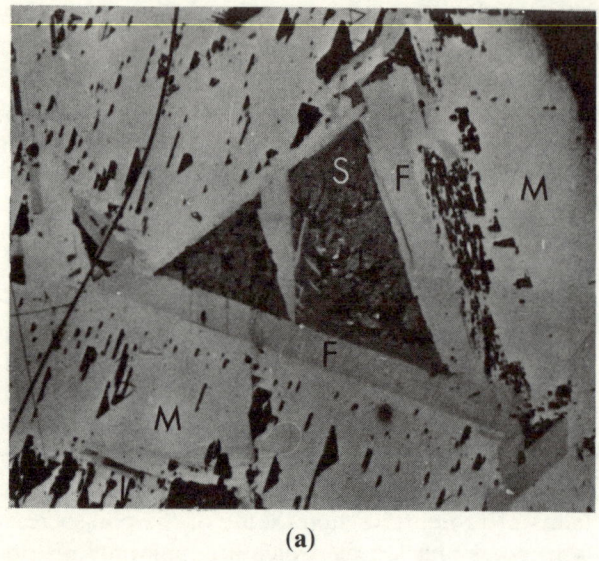

(a)

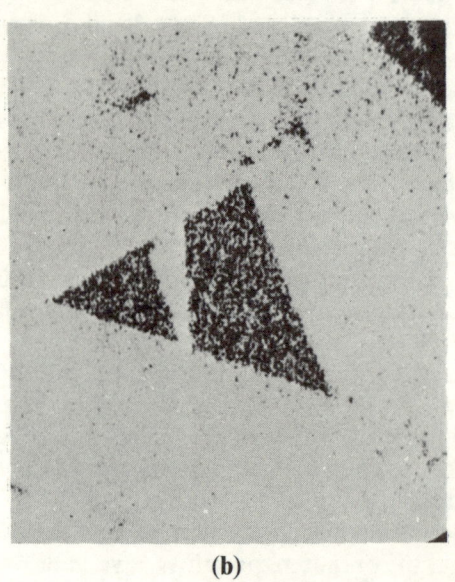

(b)

FIG. 69. (a) Optical micrograph of calcium and potassium aluminosilicate crystal (S) and potassium ferrite (F) in unreduced magnetite (M). (× 200.) (b–e) Electron probe micrographs of same area showing the distribution of iron, aluminium, calcium, and potassium. The density of white spots is proportional to the element concentration.

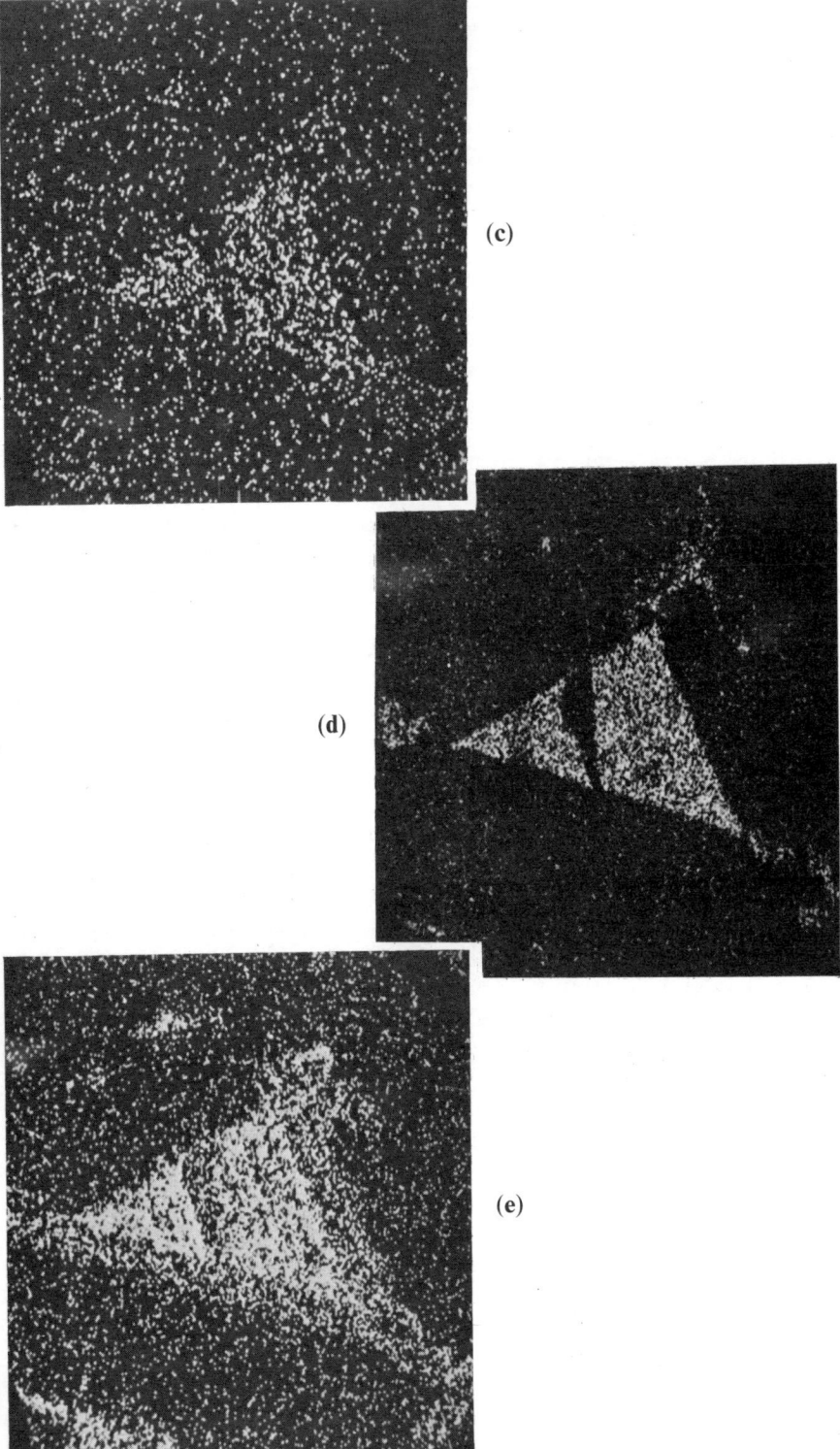

adsorption methods. These studies show that the promoters, particularly the potassium, are intimately mixed with the iron and that up to 90 per cent or more of the iron surface is covered by them, only the remaining 10 per cent or less being available for the catalysis. Nevertheless, the available iron area and the activity of the promoted catalyst are significantly greater than that of unpromoted iron, and decrease much more slowly with time.

Catalyst poisons

During the reduction process, as the reduction proceeds, the total surface area of the catalyst increases steadily, but the iron surface area increases at a faster rate towards the end of the reduction, and the activity increases correspondingly. This indicates that the catalyst surface is very heterogeneous, and that the most active iron sites are the last to be reduced, and are made available by the removal of the last oxygen atoms. The most active part of the catalyst constitutes only a small proportion, perhaps only 1–2 per cent of the total iron surface. Further evidence that the proportion of active surface is very small is provided by the sensitivity of the catalyst to gaseous poisons such as oxygen, sulphur, arsenic, phosphorus and chlorine. Very small amounts of these compounds have a drastic effect on the activity of the catalyst. Plant catalysts which have lost activity as a result of poisoning have been found on discharge to contain less than 0·1 per cent sulphur. Significant effects are produced by much smaller amounts, for example, around 0·01 per cent sulphur, which, on a new catalyst of crystallite size of 400 Å, is equivalent to less than 1 per cent of the iron surface having been sulphided.

Temporary poisoning

Oxygen and oxygen-containing compounds such as H_2O, CO, and CO_2 are well-known poisons. In the presence of a large excess of hydrogen and an active catalyst, all oxygen containing compounds are rapidly converted to H_2O, and experience has shown that these poisons are about equivalent on an oxygen basis, 100 ppm O_2 having about the same effect as 100 ppm CO_2, or 200 ppm CO or H_2O. If the catalyst is operated in the presence of the poisons for comparatively short periods the poisoning effect is temporary and reversible, so that the activity of the catalyst is restored completely by operating on pure gas.

Thus with a catalyst operating at 450°C and 300 atm pressure (space velocity 15000 m² gas/m² catalyst h), 100 ppm CO in the inlet gas reduces the ammonia concentration in the exit gas by 25 per cent in six days, the activity remaining constant at this low level while the poison concentration is maintained. Activity is, however, completely restored in one day's operation with pure gas. Gas containing 500 ppm CO reduces the exit ammonia concentration by 67 per cent in three days, and activity is completely recovered in four days' operation on pure gas. At 500°C, on the other hand, 50 ppm O_2 reduces the exit ammonia concentration by only 4 per cent but this effect is permanent, and no recovery is obtained with pure gas.

Many investigations have been made on the effect of poisons, and there is evidence that the amount of oxygen which the catalyst picks up is proportional to $\sqrt{(P_{H_2O}/P_{H_2})}$. This evidence suggests that temporary poisoning is caused by oxidation of the comparatively small proportion of highly active iron surface. The oxygen concentration which poisons the catalyst is very much less than that required thermodynamically to oxidise bulk iron to Fe_3O_4. At 450°C, P_{H_2O}/P_{H_2} must exceed 0·16 for oxidation to take place, whereas in practice significant poisoning occurs at a value of 50×10^{-6} or less, as the above figures demonstrate. It is possible that the poisoning involves the oxidation of highly active iron of much higher free energy than iron in bulk, and that the surface of the catalyst can be considered to consist of iron with a range of free energies and activities.

Permanent poisoning

When poisoning is continued for longer periods (weeks rather than days) some permanent effect is produced, and the activity of the catalyst is not completely recovered by operation on pure gas. High temperature operation accentuates the permanent effect.

In the presence of oxygen compounds, the rates of oxidation and reduction of iron are both significant. Reoxidation and re-reduction of the iron probably occur as a continuous process, during which the iron may sinter. The rate of sintering increases when the H_2O concentration increases, particularly at higher temperatures. The increase in size of the iron crystallites is not reversible, so that permanent loss of iron area and activity results.

Surface compounds are probably formed by the other well-known poisons—sulphur, arsenic and phosphorus—and they are likely to be more stable than the oxygen compounds, so that the recovery of catalyst activity after poisoning occurs very slowly, if at all. Lower concentrations of these elements (perhaps one tenth of that of oxygen) produce comparable poisoning effect. Chlorine probably causes deactivation by the mechanism described in Chapter 2, and, because KCl is volatile, also may cause loss of alkali from the catalyst.

In addition to chemical poisoning of the iron surface, loss of activity can result from physical covering of the catalyst surface and obstruction of the pores. This occurs when carbon is produced by the cracking of hydrocarbons, such as compressor lubrication oil, or by the polymerisation of olefines. Sulphur-free lubricants must obviously be used to avoid chemical poisoning of the catalyst.

Even in the absence of poisons, some recrystallisation of the iron surface occurs slowly—a process which proceeds faster at higher temperatures. In general a temperature of 540°C should be regarded as an upper limit for a catalyst such as 35–4, though this depends on the duty required. As discussed above, this crystal growth and loss of iron area is speeded up still more by the presence of oxygen-containing compounds.

Catalyst reduction

An understanding of the structure of the catalyst and of the effect of poisons underlies the manner in which the catalyst is reduced and used. During the reduction, iron produced in one part of the catalyst should not be exposed to water produced from the reduction of other parts of the catalyst. In a single particle this cannot be avoided, as iron produced on the surface of the particles is exposed to water produced by the later reduction of the interior of the particle. Because of this, larger catalyst particles tend to have lower intrinsic activity than smaller sized catalyst particles, which have been exposed less to water during reduction. (Smaller particles are also more reactive because they are less affected by gaseous diffusion, as discussed later.) In a bed of catalyst during reduction, water from the reduction of the lower (exit) parts of the bed must not come into contact with the upper (inlet) reduced catalyst as a result of back diffusion or back mixing. Water must be removed from the exit gas by cooling when it is recycled.

The effect of water as a poison is minimised by the water concentration being kept at a low level, which is ensured by use of a high space velocity (gas volume/catalyst volume), and also by reduction at as low a temperature and pressure as possible. Ammonia synthesis starts as soon as some iron is produced in the converter, and the exothermic synthesis heat raises the catalyst temperature, which increases the rate of reduction. This, therefore, must be controlled by the pressure being kept low, and by the usual temperature control methods (admission of cold gas, and so on), which decreases the make of ammonia. The rate of reduction is indicated by the quantity of water in the exit gas, which is usually not allowed to rise above 10000 ppm (vol/vol). Claims for other catalysts, possibly because of the types and distribution of promoters employed, suggest that they are more sensitive to water.

The previous sections indicated that overheating of the catalyst must be avoided at all times, since high temperatures cause some crystal growth (sintering), even with the alumina promoter present. During reduction it is desirable not to heat the catalyst above the temperature at which it is to be operated—that is, 450–500°C. The reduction, therefore, should be carried out with close control of temperature to avoid local hot spots, especially when the synthesis reaction has started. Temperature control is easier at lower pressures. The start of the reduction process, and the rate of reduction at any moment, can be determined by estimating the water content of the exit gases. The concentration should not exceed 10000 ppm (vol/vol) which is equivalent to 30 kg H_2O h^{-1}/t of catalyst at a space velocity of 10000 h^{-1} (60 kg H_2O h^{-1}/t at 20000 h^{-1}, and so on).

Pre-reduced catalyst

Under good conditions a catalyst can be reduced in about 24 hours but, when the gas rates are limited and with large converters containing 100 tons or more of catalyst, the process can take more than a week. The ammonia liquor produced during this time contains water, which can be an embarrassment for some plants. Frequently production time is lost while the catalyst is being reduced. These limitations can be minimised when a pre-reduced catalyst is used—that is, a catalyst which has been reduced before charging to the converter, usually by the catalyst supplier. In the reduced state the catalyst is pyrophoric and would become overheated by exposure to air, so that to facilitate storage, transport, and catalyst charging, the pre-reduced catalyst is stabilised (by the supplier). This is usually done by the partial re-oxidation of the reduced catalyst by its exposure to low concentrations of oxygen. The iron surface, oxidised in this way, can then be exposed to air at room temperatures without further oxidation and overheating of the catalyst. The stabilised catalyst can be charged to plant converters and re-reduced in the usual manner. The stabilised catalyst is less than 10 per cent oxidised, and on subsequent re-reduction produces less than 10 per cent of the water and weak ammonia liquor produced by the standard catalyst. The reduction procedure as a whole is shortened and the converter goes 'on the make' sooner. The use of pre-reduced catalyst can shorten reduction procedures by a factor of two or more, which often justifies its higher cost. Most advantage is gained when the whole converter charge is pre-reduced, but pre-reduced catalyst can be used as a part-charge at the top (inlet) of a converter, where it will quickly initiate the synthesis reaction, and provide a source of heat to supplement the converter heater. This is beneficial as it enables a higher gas space velocity to be used during reduction.

The pre-reduction and stabilisation is carried out by the catalyst supplier. Very careful control of the process is essential, otherwise some loss of activity results when the catalyst is re-reduced. Pre-reduced ICI catalyst 35–4, when compared with normal catalyst, shows no loss of activity in plant converters. It has to be stored carefully in sealed drums in a dry, cool place, otherwise some further re-oxidation and overheating can occur.

Kinetics of ammonia synthesis

Despite the fact that ammonia has been successfully synthesised commercially for more than 50 years the detailed mechanism of the reaction is still a matter of some discussion.

A high rate of reaction is favoured by high temperatures and pressures, but high temperature implies a lower equilibrium value of ammonia concentration, and hence smaller 'driving force'. Hence the rate of reaction increases as the temperature is raised, but reaches a maximum value and then falls as equilibrium is approached, so the optimum yield at a given pressure is obtained with the temperature profile falling along the catalyst bed as the ammonia percentage increases. Under commercial conditions the maximum reaction rate is at a temperature about 70°C below the equilibrium temperature. A rate expression to describe ammonia synthesis thus has to take account of temperature, pressure, gas composition and the equilibrium composition.

Many reaction mechanisms have been proposed but, because of their almost universal application to plant design, only those of Temkin and his co-workers are considered here. The first kinetic equation to give reasonable agreement with observed rates was that due to Temkin and Pyzhev (ref. 70) in 1940. This equation was based upon the assumption that the adsorption of nitrogen on a non-uniform surface is the rate controlling step, and it led to the now well known equation for the intrinsic reaction rate, which is that, in the absence of diffusion,

$$w = k_2 \left\{ K_p P_{N_2} \left(\frac{P_{H_2}^3}{P_{NH_3}^2} \right)^\alpha - \left(\frac{P_{NH_3}^2}{P_{H_2}^3} \right)^{1-\alpha} \right\}, \tag{1}$$

where w = rate of reaction
K_p = equilibrium constant for $N_2 + 3H_2 \rightleftharpoons 2NH_3$,

the constant α has a value between 0 and +1, and

$$k_2 = k_{2(0)} \exp-\left\{\frac{\Delta E_{k_2}}{R}\left(\frac{1}{T}-\frac{1}{T_0}\right)\right\}$$

$\Delta E_{k_2} \approx 38.00$ kcal/mol.

This equation has been the basis of industrial converter design for the last 20 years. Most workers, including those at ICI (ref. 71) use the value of α found by Temkin (that is, α = 0.5). Others, notably Nielsen (ref. 69), have found their results best supported a figure of 0.75. In general it has been found necessary to allow k_2 to decrease with increasing pressure, although again Nielsen (using α = 0.75) and Livshits and Sidorov (ref. 72) (using α = 0.5) claim that k_2 is substantially pressure-independent if fugacities rather than partial pressures are used to allow for non-ideality. Inspection of equation (1) shows that it cannot apply when the ammonia concentration is zero, since it then predicts infinite reaction rate. More recent work (ref. 74) has established that under these conditions the rate is given by

$$w = kP_{H_2}^\alpha P_{N_2}^{1-\alpha} \qquad (2)$$

In 1963 Temkin, Morozov, and Shapatina (ref. 73) proposed a mechanism which incorporated, as an important step, the addition of the first hydrogen molecule to the adsorbed nitrogen. They obtained

$$w = \frac{k_* P_{N_2}^{1-\alpha}\left\{1-\dfrac{P_{NH_3}^2}{K_p P_{N_2} P_{H_2}^3}\right\}}{\left\{\dfrac{l}{P_{H_2}}+\dfrac{1}{K_p}\cdot\dfrac{P_{NH_3}^2}{P_{N_2} P_{H_2}^3}\right\}^\alpha\left\{1+\dfrac{l}{P_{H_2}}\right\}^{1-\alpha}} \qquad (3)$$

with

$$k_* = k_{*(0)} \exp-\left\{\frac{\Delta E_{k_*}}{R}\left(\frac{1}{T}-\frac{1}{T_0}\right)\right\}$$

$$l = l_{(0)} \exp-\left\{\frac{\Delta E_l}{R}\left(\frac{1}{T}-\frac{1}{T_0}\right)\right\}$$

It can be shown that under the two extreme conditions of close to and far from equilibrium equation (3) becomes:

(i) Equation (1) with $k_2 = k_*/K_p^{(1-\alpha)}$ \qquad (4)

(ii) Equation (2) with $k = k_*/l^\alpha$ \qquad (5)

If k_* is truly pressure independent, then equation (4) suggests the pressure dependence of k_2 through the factor $K_p^{(\alpha-1)}$.

In recent years there has been a considerable amount of work on ammonia synthesis in the ICI laboratories, and catalysts have been tested in both integral and differential types of reactors, which are described in Chapter 3. It has been found, from a large number of differential rate determinations over a wide range of conditions, that this model of Temkin, Morozov and Shapatina [equation (3)] gives, for non-diffusion limited rates, a much better fit than equation (1). Under most conditions of commercial interest equation (3) corresponds quite closely with equation (1) with $k_2 = k_*/K_p^{(1-\alpha)}$. The 'best fit' values, $\Delta E_{k_*} = 26.455$ kcal/mol and $\alpha = 0.465$, agree quite well with those of Temkin et al. (ref. 74) ($\Delta E_{k_*} = 25.0$ kcal/mol, and $\alpha = 0.5$), and k_* was found to be independent of pressure. Thus it is possible to calculate the pressure and temperature dependence of k_2. At 200 atm and 450°C, K_p varies as $P^{0.44}$; and if $\alpha = 0.465$, k_2 varies as $P^{-0.24}$. At low pressures, K_p is a function of temperature only and k_2 then has no pressure-dependence. Similarly the activation energy of k_2 can be predicted as being:

$$\Delta E_{k_2} = \Delta E_{k_*} + (1-\alpha)\Delta H_R$$

where ΔH_R = heat of reaction

$$= -RT^2\left(\frac{\partial \ln K_p}{\partial T}\right)_p \approx 26 \text{ kcal.}$$

Hence $\Delta E_{k_2} = 26.5 + 13.9 = 40.4$ kcals, which is close to the normally quoted value of 38 kcal/mol.

Effect of catalyst size

Kinetic expressions so far discussed have been derived for conditions where the rate of reaction is not limited by diffusion. Rate measurements made with the catalyst in a variety of size ranges show that diffusion has a marked effect, particularly at high temperature. This is illustrated in fig. 70 where reaction

	Rate (kmol $N_2 h^{-1}$ per m^3 catalyst)
Crushed 35–4*	300
$\frac{1}{8}-\frac{3}{16}$ in	112
$\frac{1}{4}-\frac{3}{8}$ in	61

*Size 14–25 BSS. (0.600–1.205 mm)

FIG. 70. Variation of synthesis rates with granule size.

rates for crushed, $\frac{1}{8}-\frac{3}{16}$ in, and $\frac{1}{4}-\frac{3}{8}$ in catalyst are compared. The measurements were made in a differential reactor at 500°C and 100 atm with a 3:1 $H_2:N_2$ gas mixture containing 4 per cent NH_3.

It is clear that the larger size catalyst particles are considerably less active than the smaller particles. This is in part the effect of diffusion limitations within the catalyst pores. Another contributary factor is the lower intrinsic activity of the reduced iron surface in larger catalyst particles, due to the outer layers of iron of large particles being exposed to the poisoning action of water (see p. 139) produced by reduction of magnetite within the particles.

Catalyst of the particle sizes used commercially are clearly subject to both these effects, and their magnitudes are dependent upon gas composition, temperature, and pressure, as well as upon catalyst composition and structure. Measurements have been made on catalyst 35–4 with the test equipment described below, which have permitted the kinetic equation (3) to be modified to take into account both diffusion and the effect of particle size when commercial converters are being designed. The reaction rates in the table above are for fresh catalyst 35–4, and are not suitable for sizing a converter, for which purpose the relative rates for the aged catalyst are necessary. Intrinsic activity declines during use, due to poisons and sintering. The extent of this decline is very dependent upon the operating conditions, and the purity of the synthesis gas, so appropriate information must be available when the design activity is calculated.

Ammonia synthesis catalyst testing

There are two distinct reasons for catalyst testing. The development of kinetic equations requires accurate information which is difficult to obtain, whereas the evolution of new catalysts requires rapid, though not particularly accurate,

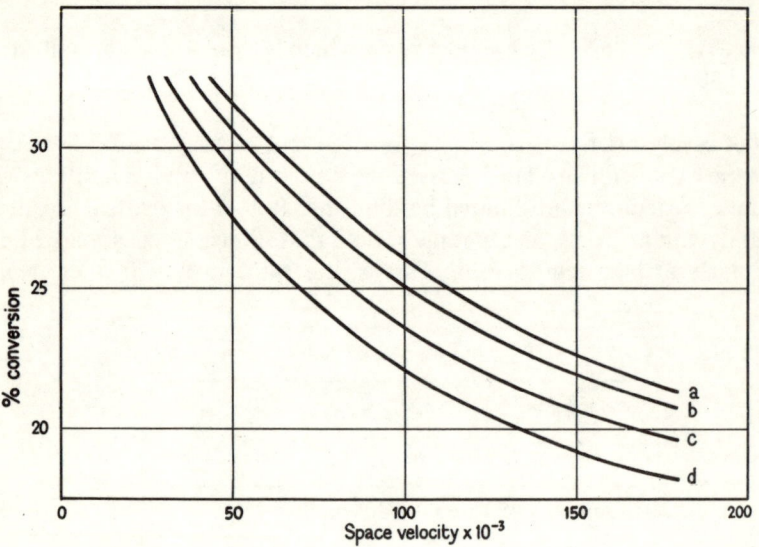

FIG. 71. Relative activities of ammonia synthesis catalysts. Per cent conversion against space velocity for integral, isothermal reactors. Pressure, 200 atm; temperature, 450°C; inlet ammonia, 0 per cent. All samples crushed to 14–25 BSS before testing (consequently synthesis not affected by diffusion).

sorting techniques. There are different types of activity testing apparatus which meet these requirements.

The isothermal integral reactor is commonly used in studying synthesis kinetics, but its accuracy leaves something to be desired. The catalyst is contained in an electrically heated tube maintained at constant temperature. Ammonia is synthesised over a range of temperatures, pressures and space velocities, and the overall (that is, the integral) conversions are interpreted in terms of a kinetic rate equation such as equations (1)–(3). The considerable reaction heat is dissipated either longitudinally or through the reactor walls in an attempt to maintain, as closely as possible, a uniform temperature profile through the catalyst bed. Whilst this ideal can be approached quite closely for crushed catalyst of equivalent diameter less than 1 mm, it becomes more difficult as the catalyst size is increased, and is virtually impossible for, say, 6–10 mm catalyst.

Such reactors are relatively simple to operate and are therefore well suited to routine catalyst testing, and as a means of comparing the intrinsic activities of new catalyst formulations. Fig. 71 shows typical results for a series of catalysts. The space velocities at which a given ammonia concentration is obtained are a measure of catalyst activities. A comparison between these curves shows that catalyst 35–4 has good intrinsic activity (curve a).

Great precision cannot be ascribed to the absolute activities calculable from these curves, for the reasons mentioned earlier, but they are a good qualitative measure, and show the isothermal integral reactor to be an ideal 'catalyst-sorting' unit, as well as a reasonable quality testing reactor. It has also been used to study the effect of gaseous poisons on the catalyst.

The best type of reactor for studying the four prime variables (temperature, pressure, gas composition and catalyst size) is a differential reactor. A recycle reactor (in which gas is recirculated around a loop containing the catalyst at a much higher rate than the make-up gas and take-off gas rates) allows reaction rates to be measured at a known pressure, temperature, and (essentially constant) gas composition. This type of reactor was first suggested in the study of ammonia synthesis by Temkin *et al.* in 1950 (ref. 75). ICI operates such a reactor capable of working over the following range of conditions:

Temperature	up to 550°C
Pressure	50–350 atm
Ammonia percentage	0–16 per cent
Catalyst size	0·5–15 mm

The recycle reactor enables differential rate determinations to be made over this whole range of conditions on catalysts of different sizes, and hence the catalyst effectiveness can be measured. This reactor has been used extensively by ICI to evaluate their own catalysts and, as indicated previously, to develop kinetic equations for converter design purposes.

Conclusion

The ammonia synthesis catalyst has had almost the longest history of full-scale use and development of any catalyst. It led the way to high-pressure catalytic processes and thereby opened up much of the chemical industry which we take for granted today, from methanol synthesis to petroleum refining. Research on the catalyst itself, which has been both empirical and fundamental, revealed for the first time many aspects of catalysis—for example, the function of promoters and activators—which are common knowledge today.

After reading this brief account of the synthesis catalyst, it might be thought that 50 years of extensive research and development has brought the catalyst to its ultimate state of development—to some equilibrium state controlled by economic and scientific factors. These factors, however, are not constant. New raw materials (oil rather than coal) produce cheaper synthesis gas than earlier processes. The increase in size of synthesis units and the availability of improved gas compressors can influence the pressure of the synthesis process. New operating conditions require new catalysts. In addition to these economic factors, limitations of the present catalysts in respect of activity and life and poisoning become apparent. Operation at even lower temperatures and pressures would have obvious advantages.

Consequently there is a continuing incentive, particularly in industrial laboratories, to increase knowledge in this field and to improve the catalyst.

CHAPTER 8

Computer programs for converter calculations

THERE are basically three types of question which can be asked about a converter. These are:

(a) What size (and possibly what kind) of converter is needed for a particular task?
(b) Given a particular converter, how will it behave under assigned operating conditions?
(c) Given both a particular converter, and its behaviour under some set of operating conditions, what is the activity of the catalyst?

The three classes of problem to which these questions give rise may be termed those of design, of performance and of simulation.

In a design problem it is assumed that a specific task exists—for example, the reduction from 7·5 per cent to 0·3 per cent of the CO concentration of a particular gas stream—and that the properties (primarily, the activities) of the catalyst or catalysts which may be used are known. The aim is then to calculate the volumes and configuration of the appropriate catalyst beds. Commonly, the wish will be to make these as cheap or as small as possible.

In a performance problem, the specification of the converter is known, and the aim is to predict its behaviour under stated operating conditions. Where conditions may be varied, the calculation is normally required to indicate those which will produce the best results.

In a simulation problem, both the specification of the converter, and its performance under given conditions, are known, and the purpose of the calculation is to deduce the catalyst activity which these imply. Here no question of seeking the best solution arises, for all the conditions are fixed. We may indeed have over-specification of the problem, with more data to hand than are strictly necessary for the definition of a solution. In such circumstances we may be forced to a sort of 'inside-out' optimisation in the form of a least-squares fit to the more or less inconsistent data.

The separate problems will often be associated. For example, we may wish to calculate the performance under a range of conditions of a converter we have just designed for some specific duty, or we may wish to optimise the performance of a converter which we have just simulated. Thus the three types are not wholly distinct. In particular, they all require for their solution a knowledge of the kinetics and thermodynamics of the reaction or reactions involved, and an ability to calculate therefrom the behaviour of a given mass or volume of catalyst under specified operating conditions. It is at this point that the advantages of digital computers become manifest. The kinetics of a catalytic reaction are seldom representable by a rate equation simple enough to be integrated analytically, while the difficulties are customarily compounded by the thermal effects of reaction, or of any applied cooling. Thus calculation of the performance of a catalyst bed must in general be effected by numerical integration of one or more simultaneous differential equations—a tedious task by hand and one well suited to automatic digital computation. Not without reason, therefore, converter programs were among the first to be written when, in the mid-1950s, high-speed digital machines began increasingly to become commercially available.

The first converter programs were simple performance programs which would give, for any completely specified system, the result—for example, exit product concentration—to be expected. Their primary aim was, as already indicated, to replace the equivalent and time-consuming hand calculations which had previously been necessary. By reducing, however, to something under a five minutes' task what had earlier taken perhaps two days of hand computation, they immediately opened up the prospect of a much more systematic optimisation, either of design or of performance, than had hitherto been practicable.

Optimisation

The opportunity—indeed the need—for optimisation arises whenever a problem is incompletely defined—when, for example, the only requirement set for a design is that the converter be capable of effecting a specified conversion of a specified feed stream. With such an incomplete formulation, an infinitude of solutions is possible, each meeting the prime requirement but differing in detail, and definiteness is only restored to the problem by requiring also that the acceptable solution is that which is 'best' in some way—that is to say, is optimised with respect to those aspects not overtly specified, and hence at the discretion of the problem solver.

To the designer, optimisation is synonymous with cost minimisation. Thus, the optimal converter is that which will yield the specified conversion at minimal cost. So far as kinetic aspects are concerned, this reduces to the minimisation of the necessary catalyst volume or (where more than one catalyst is involved) the catalyst cost. In practice, such minimisation may need to be carried out for each of a small number of distinct configurations leading to intrinsically different vessel costs—for example, in ammonia synthesis we may wish to cost the optimal three-bed converter, the optimal four-bed converter, and the optimal tube-cooled converter. This, however, is a matter of expediency rather than principle.

To the plant operator, optimisation implies maximising profits. With the converter type and size fixed, the aim of optimisation must then be to choose from the available sets of operating conditions (temperatures, flow rates, and so on) that set which will do just this—the one that will lead to most effective use of the converter. The kinetic problem is again somewhat narrower—it is that of choosing, for a fixed converter and feed stream, that set of operating temperatures and/or cold shot and by-pass rates which will give maximum conversion. Variations in flow rate or in feed composition will introduce cost considerations extraneous to any kinetic aspects, so that in practice, optimisation may need once more to be carried out for a number of distinct cases, which can later be costed and compared. The main computational burden nevertheless remains that of deriving for any assigned inlet conditions the maximum achievable conversion and the means whereby this may be achieved.

The greater facilities which digital computers offered were eagerly exploited, and it rapidly became not uncommon for as many as a hundred calculations to be run in the course of a single design study, rather than the two or three which were all that could previously be envisaged. Concurrently, the awareness grew that, with the possibility of such a massive attack on any problem, there was no longer the need to rely on the skill and accumulated experience of the designer—commodities inevitably and always in short supply. To some extent, indeed, such reliance was no longer a good thing, carrying as it did the risk of perpetuating the errors and misconceptions of the past. There was distinct advantage in leaving the whole task of optimisation to the computer, and thus, as it were, approaching each new problem with an open mind. From such thinking two lines of development emerged—the application, to converter problems, of the general numerical optimisation techniques which were coming forward in increasing numbers, and the establishment—more particularly for multi-bed adiabatic converters—of specialised optimisation techniques, making as much use as possible of underlying analytical relationships to minimise the computational effort. Both approaches have been utilised in the development of specific converter programs within ICI.

The generalised numerical optimisation techniques, of which there are now many, are extensively described in the computer literature, and need not be discussed in detail here. Essentially, they depend on the execution of an orderly sequence of calculations in which earlier results are used to indicate those later calculations likely to lead to more favourable outcomes. That is to say, attention is focussed on 'hill-climbing' rather than on mapping of the whole region of possible interest. For optimal design of tube-cooled ammonia converters, use is made of the Simplex technique (ref. 76), which is well adapted to the essentially two-dimensional problem posed. The sequence of calculations, when plotted against bed inlet temperature and bed cooling factor (the two variables at the designer's discretion), then takes the form of a chain of contiguous triangles (two-dimensional simplices) stretching towards and eventually circling the point representing optimum conditions. Final location of the optimum is facilitated by the fitting of a quadratic approximation to the terminal hexagonal design.

Of more importance are the techniques used for optimal design of multi-bed converters—that is, shift converters, or ammonia cold-shot converters. Such problems are not so readily treated by general techniques, because of the greater number of variables involved and, more particularly, the constraints on operating temperatures which are customarily applied. The essential theory of the optimal design of such reactors is given by Horn (ref. 77), together with an elegant procedure for the treatment of working temperature constraints. Horn considered, however, only the case of a converter with indirect (heat exchanger) cooling between beds. Our own work has shown that an analogous treatment is possible for any form of interbed cooling, and has extended the treatment to converters with mixed cooling arrangements, such as are common in shift conversion. Procedures have also been developed for the automatic handling of overall temperature rise constraints, which occur in the design of those converters intended for autothermic operation (in ammonia synthesis, for example).

The theory turns on the existence of two types of condition applying to any optimal design—the one (a variational condition) applying to each bed in turn, and the other (a 'rate match') applying to the extent of cooling between each bed and the next. These conditions are modified in an explicitly calculable manner by the incursion of working temperature constraints. By utilising the conditions, the determination of an optimal design may be reduced to a search in a number of dimensions which does not exceed the number of distinct types of cooling used. This is in contrast to the $2n-1$ dimensions of search which must be explored (for an n-bed converter) by any general numerical optimisation technique which cannot utilise the underlying theory.

For optimal performance calculations, the theory is of less help, since the fixing of each bed volume destroys the $(n-1)$ 'rate match' conditions, and a search is necessary for each of the n bed inlet conditions. The systematic procedure which the theory indicates is still, however, preferable to the use of a generalised search technique. The searches for each bed inlet condition are in the main sequential rather than nested, so that the amount of computation required increases linearly rather than exponentially with increase in the number of beds.

For multi-bed converters, the theory thus gives rise to two distinct classes of program—those for optimal design, and those for performance optimisation—which replace in their separate areas the 'cut-and-try' methods necessary with the earlier performance programs, and which still underlie the use of generalised search techniques. For the tube-cooled converter, no such separation is necessary. By taking as the quantity to be optimised the ratio (exit product concentration)/(catalyst volume), and terminating integration either when (in design) the desired product concentration is reached or when (in a performance optimisation) the available catalyst volume has been traversed, the same program may be used for both design and performance optimisation.

Requirements of an effective program

It goes without saying that an effective program, whether for design or for performance optimisation, must incorporate an effective optimisation technique. Beyond this, however, there are a number of requirements, essentially of convenience, which it should satisfy. The data input should be as simple as possible; prefatory calculations or conversion of units should not be necessary and, where code entries must be used (to indicate particular types of inter-bed cooling, for example), confusion must be avoided. The submission of data is facilitated by the use of standard data forms of the question and answer type, which are designed to elicit no more than an exact specification of the problem, and which clearly indicate any coding to be used. The lay-out of the form should be chosen to facilitate the punching of the data on to card, to minimise the risk of transcription errors. For the same reason the data format for card reading should preferably not be too critically dependent on exact alignment of the entries in particular columns.

The commoner data faults—missing items, items out of scale, percentages failing to add to 100—should be detected by the program, and an informative message output. To safeguard the integrity of later data sets, no attempt should be made to execute a calculation before such checks are made. Standard items (of thermodynamic or kinetic data, for example) should be embedded within the program, and should not need to be specified at run time. This is particularly important where the alternative involves reference to tables or graphs, and misreading is an inherent possibility. Nothing will destroy faith in a computer program faster than inconsistency in its results.

The programs should be (externally) as flexible and as general as is reasonably possible. 'Special cases' should be detected within the program and be dealt with appropriately, so that it is not incumbent on the user to be able to recognise such special cases. There is, nevertheless, a limit to the flexibility and generality which can be built in—determined not so much by conceptual difficulties as by considerations of cost-effectiveness and ease of use—and this limit should not be over-stepped. Thus, for example, while there is advantage in dealing with both high- and low-temperature shift converters by means of a single program, because of the great area of common ground and the frequent practical use of both in combination, there would be no virtue in combining ammonia and shift converter design within a single package—this would simply be wasteful of core storage and central processor time.

At the final stage, the output from the program should be comprehensive and readily assimilable, with adequate annotation and—in particular—full details of the input data. This last is very necessary to guard against the possibility that, because of some error of specification, the wrong calculation has been run. The aim should be to produce a complete record of the calculation, capable of being filed or passed on without the need for transcription. It is pointless to take great pains to guard against transcription errors in the input data, if similar steps are not taken to safeguard the results. Where multiple copies are required, it is advantageous to arrange for these to be produced directly by the program, since this will permit, also, variation of presentation to meet the needs of different recipients.

In the sections which follow, some of the more important programs utilised for converter design within ICI Agricultural Division are briefly described and illustrated, to illuminate some of the points made above. They are available for the handling of customers' enquiries on catalyst matters, and incorporate at any time the best available current information on the kinetics and thermodynamics of the reactions concerned.

Shift converter programs

RTC00 program

The basic shift converter optimal design program is that designated RTC00. A specimen data form and computer output for this program are shown in fig. 72. The program is designed to accept data and to output results in

152 COMPUTER PROGRAMS FOR CONVERTER CALCULATIONS

OPTIMAL DESIGN

RT/PL/17

Program: RTC00
OS: 4
P
Time (mins)
Job No.

A. Converter Data

Arrangement of catalyst beds					
Interbed cooling : bed 1/2					
(if quench, state form & temp. (°C)) bed 2/3					
Sulphur concentration in dry gas (p.p.m.)					

B. Basic Data

Basic Data		Units	Computer Data				
Title			TEST (ARTC00)				
Date							
* Th. Exp. Ref. No.							
Number of beds			2.				
* c			2.				
Inlet	dry CO	%	12.49				
	H₂	%	56.71				
	CO₂	%	7.71				
	N₂	%	22.64				
	CH₄	%	0.45				
Steam to dry gas ratio			0.5644				
Pressure		ata	28.				
Exit dry CO or exit conversion		%	0.30				
Inlet dry gas rate		Rm³/hr	130488.				
* To		°C					
* Tq		°C					
Catalyst cost index	bed 1		1.				
	bed 2		5.				
	bed 3						
* Mc			-2.				

C. Overwrites

Overwrites		Units	Item No.	Additional Computer Data			
Temperature inlet	bed 1	°C	1	390.			
	bed 2	°C	2				
	bed 3	°C	3				
Conversion of gas inlet 1st bed		%	5				
Activity of catalyst	H.T.	fr.	6				
	L.T.	fr.	7				
Max. working temp.	bed 1	°C	8				
	bed 2	°C	9				
	bed 3	°C	10				
Min. working temp.	bed 1	°C	12				
	bed 2	°C	13				
	bed 3	°C	14				
* Dc				-1.			

```
                I.C.I. WATER-GAS SHIFT CONVERTER DESIGN        RT0


    CLIENT        TEST (ARTC00)
    DATE
    I.C.I. REF.NO. W.G.S.


    INLET DRY GAS COMPOSITION   CO(PER CENT) 12.49
                                H2            56.71
                                CO2            7.71
                                N2            22.64
                                CH4            0.45

    STEAM/DRY GAS RATIO                        0.564
    AVERAGE PRESSUPE(ATAS)                    28.00
    REQUIRED EXIT DRY CO(PER CENT)             0.30
    DRY GAS RATE(RM3/HR)                  130488

    CONVERSION(PER CENT)                       0.00

    BED 1 MAX INLET TEMP(DEG.C)              540.00

                              L.T.S.     H.T.S.
    ACTIVITY      PRESET       2.88       2.37
                  OVERWRITE     -          -


                              1ST BED    2ND BED

    T MAX(DEG.C)  PRESET      540.00     250.00
                  OVERWRITE     -          -

    T MIN(DEG.C)  PRESET      330.00     200.00
                  OVERWRITE     -          -

    T INLET(DEG.C) PRESET     OPT.       OPT.
                   OVERWRITE  390.00      -


    CATALYST                  15-2       52-1
    CATALYST VOLUME(M3)       33.09      73.98
    CATALYST COST INDEX        1.00       5.00
    TEMP(DEG.C)   INLET      390.00     216.84
                  EXIT       454.20     241.02
    EXIT DRY CO(PER CENT)      3.20       0.30

    CONVERSION(PER CENT)      72.09      97.31
    T EQUILIBRIUM(DEG.C)     460.90     246.49
    ADIABATIC EQ.(DEG.C)     455.09     241.24
    "  "    (%DRY CO)          3.08       0.27

    INTERBED COOLING
      - /BED 1                HEAT EXCHANGER
    BED 1/BED 2               HEAT EXCHANGER
```

FIG. 72. Computer form— Water-gas shift converter/design.

metric/centigrade units, with gas rates expressed as $\overline{R}M^3/hr$—that is, as equivalent volumetric flow rates at 20°C and 1 atm. An identical program for data in Imperial/Fahrenheit units is designated RTC01.

The program will carry out calculations for either high- or low-temperature shift beds, or for a combined system, up to a maximum of three beds. For a combined system, it will take as optimal that arrangement which minimises total catalyst cost, the ratio of catalyst costs (LT/HT) forming one item of input data. Inter-bed cooling may be by indirect heat exchange, by steam quench, or by condensate quench. Mixed arrangements in which two forms of cooling are used may be handled automatically, though it is assumed that (as is customary in practice) the first bed is preceded by a heat exchanger only.

Standard working temperature constraints are built into the program in the form of preset parameters, which may be over-written by the user, but are reset automatically on completion of a calculation. Inlet temperatures to the first and/or last bed may be fixed, if desired, and a restricted optimum found. The program does not, however, in its present form, allow the inlet temperature to the second bed of a three-bed system to be fixed. Standard catalyst activities for ICI catalysts 15–4 and 52–1 are present within the program, but these may also be overwritten by the user if desired.

The output is—for ease of filing—arranged to occur wholly on the left half of the line-printer page. The commonest problem treated is the three-bed system—two high-temperature beds followed by a low-temperature bed—which typically requires about one minute of IBM 360/50 time.

RTC02 program

The optimal performance shift converter program (in metric/Centigrade units) is designated RTC02. Again there is an equivalent program (RTC03) for Imperial/Fahrenheit units. A data form and specimen output for RTC02 are shown in fig. 73.

The program may be used for systems with up to three catalyst beds, whether high or low temperature, and will seek to maximise conversion by choice of those bed inlet temperatures—any or all—which are not fixed by the user. Any form of cooling (heat exchange, steam quench, or condensate quench) may precede any bed, though for quench systems the tacit assumption is made that there are no restrictions on the amount of quench available which would preclude otherwise admissible inlet temperatures.

The standard working temperature constraints are preset within the program, and may again be overwritten at the user's discretion. Wherever possible, the program will seek not to violate the imposed constraints. Since, however, the catalyst volumes are fixed, circumstances may arise in which it is impossible not to violate a constraint. The program will then, for the bed or beds in question, choose the lowest permissible inlet temperature or temperatures to minimise the extent of constraint violation.

As for the design program, the output is restricted to the left-hand half of the page. Typical running times on an IBM 360/50 computer are around 10 seconds per bed.

154 COMPUTER PROGRAMS FOR CONVERTER CALCULATIONS

Converter Data / NOT Computer Data

Converter Data		
Arrangement of catalyst beds		
Interbed cooling :	bed 1/2	
(if quench state form & temp (°C))	bed 2/3	
Sulphur concentration in dry gas	(p.p.m.)	

Program		RTCO2
OS		4
		P
Time (mins)		
Job No.		

OPTIMAL PERFORMANCE

RT/PL/18

Basic Data / Computer Data

Basic Data		Units	Computer Data
Title			TEST 4 (ARTCO2)
Date			
* Th. Exp. Ref. No.			
No. of beds		.	2.
Inlet	dry CO	%	10.13
	H_2	%	57.48
	CO_2	%	9.72
	N_2	%	22.33
	CH_4	%	0.34
Steam to dry gas ratio			0.754
pressure		ata.	19.5
Inlet dry gas rate		Nm^3/hr.	75694.
* To		°C	
Bed 1	Cooling control number		2.
	Inlet temperature	°C	+1.
	Quench temperature	°C	
	Catalyst volume	M^3	37.
	Catalyst index		1.
Bed 2	Cooling control number		2.
	Inlet temperature	°C	+1.
	Quench temperature	°C	
	Catalyst volume	M^3	44.
	Catalyst index		2.
Bed 3	Cooling control number		
	Inlet temperature	°C	
	Quench temperature	°C	
	Catalyst volume	M^3	
	Catalyst index		
* Mc			-2.

Overwrites

Overwrites		Units	Item No.	Additional Computer Data
Max. working temperature	bed 1	°C	15	
	bed 2	°C	25	
	bed 3	°C	35	
Min. working temperature	bed 1	°C	16	340.
	bed 2	°C	26	
	bed 3	°C	36	
*D				-1.

```
I.C.I.  WATER GAS SHIFT PERFORMANCE


CLIENT   TEST 4(ARTCO2)
DATE
REF.NO.



          BED 1
          CATALYST      15-2
          CATALYST VOLUME     37.00 M3
          CATALYST ACTIVITY   2.37
          AVERAGE PRESSURE    19.50 ATMOSPHERES
             INLET CONDITIONS         EXIT CONDITIONS
STEAM RATIO        0.754                     0.624
CO                10.13                      1.94 %
H2                57.48                     60.64
CO2                9.72                     16.43
N2                22.33                     20.67
CH4                0.34                      0.31
TEMPERATURE      369.27                    420.19  DEG C
DRY GAS RATE   75694.                    81772. RM3/HR
(OPTIMUM INLET TEMPERATURE)

INTER BED COOLING BY    HEAT EXCHANGER
EQUILIBRIUM TEMPERATURE    436.52 DEG C




          BED 2
          CATALYST      52-1
          CATALYST VOLUME     44.00 M3
          CATALYST ACTIVITY   2.88
          AVERAGE PRESSURE    19.50 ATMOSPHERES
             INLET CONDITIONS         EXIT CONDITIONS
STEAM RATIO        0.624                     0.596
CO                 1.94                      0.18 %
H2                60.64                     61.32
CO2               16.43                     17.88
N2                20.67                     20.31
CH4                0.31                      0.31
TEMPERATURE      221.29                    234.19  DEG C
DRY GAS RATE   81772.                    83213. RM3/HR
(OPTIMUM INLET TEMPERATURE)
INTER BED COOLING BY    HEAT EXCHANGER
EQUILIBRIUM TEMPERATURE    240.27 DEG C
```

FIG. 73. Computer form— Water-gas shift converter/performance.

Conclusion

The programs have been found capable in practice of dealing automatically with over 95 per cent of the problems submitted—which by now total some thousands. They retain, moreover, sufficient flexibility to facilitate the piecemeal handling of those occasional problems—usually heavily constrained—which cannot be treated directly. Whilst there exist other programs for shift converter calculations (RTCP1 for a straight-through performance calculation, for example) these are seldom used in practice, and thus need not be described here.

Ammonia converter programs

RTK25 program

The ammonia converter programs fall into two classes—those for cold-shot converters and those for tube-cooled converters. The cold-shot converter optimal design program—designated RTK25—is broadly similar in conception to the shift-converter program RTC00—that is to say, it uses the same general theory, but it differs in a number of respects associated with the differing circumstances in which the programs and the theory must be applied. An obvious difference in the data input is the absence of any code numbers to indicate the forms of inter-bed cooling, since only the one form (cold shot) is possible. An addition to the input data (and an important one) is the minimum overall temperature rise required. The need in most instances to demand a minimum overall temperature rise arises from the fact that ammonia converters are preferably operated autothermally, and for this to be possible it is necessary that the gases on exit from the final catalyst bed be at a temperature higher than that at which they entered the first bed. Instances are not unknown where an unconstrained optimal design has violated this condition.

The operation of the program is such that, if an unconstrained optimal design will give at least the minimum required overall temperature rise, the constraint is ignored—there is no 'de-optimisation' to achieve exactly the stated figure. If, however, the unconstrained optimal design cannot achieve the stated temperature rise, the program will calculate that design having minimum volume which will just achieve this figure. If the minimum required overall temperaure rise is set to zero, the program will, in fact, calculate an unconstrained optimum.

To facilitate routine use, certain alternative input data conventions are allowed. For example, the required duty may be expressed either as an exit ammonia concentration in per cent, or as a daily make of ammonia in tonnes—provided the latter exceeds 100 it will be automatically distinguished as such, and appropriate conversions will be effected. In the same way, the catalyst activity may be defined by giving, in addition to the catalyst size, either its age and the poisons level (oxygenated compounds) in ppm in the make-up gas, or the percentage of 'new catalyst' activity to be assumed. The alternative entries occupy the same positions on the data form, and in the data string read into the computer. In this way there are no blank entries which can lead to errors in transcription and punching.

A specimen data form and computer output are shown in fig. 74. Data and

156 COMPUTER PROGRAMS FOR CONVERTER CALCULATIONS

RT/PL/15

OPTIMAL DESIGN

Program	RTK 25
O.S.	4
	P
Time (mins)	
Job No.	

		Units	Computer Data			
Title			TEST 2 (ARTK25)			
Date						
Th.Exp.Ref.No.						
20	Number of copies		2.			
21	Number of beds		2.			
22	Inlet : H_2	%	63.33			
23	N_2	%	20.994			
24	CH_4	%	10.847			
25	A	%	2.753			
26	He	%	0.			
27	NH_3	%	2.076			
28	Pressure	ata	140.			
29	Exit NH_3	%	868.			
30	Inlet gas rate	Rm^3/hour	576441.			
31	Minimum overall temperature rise	°C	60.			
32	Cold shot temperature	°C	142.			
33	Catalyst size	mm.	6.			
34	Catalyst age	years	67.2			
35	Impurities in M.U.G. (eg. O_2 Cpds.)	p.p.m.				
	Control number		-2. xx			
			1.			
			400.			
			3.			
			400.			
			5.			
			400.			
			-1.			

ADDITIONAL RELEVANT INFORMATION:

```
            NH3 COLD SHOT CONVERTERS - OPTIMAL DESIGN        COPY    1
TEST 2(ARTK25)
                                                             NH3
PROGRAM CONSTANTS
   MIN T INLET BED 1           400.0      MAX T FOR RATE EQUN   610.0
   MAX T INLET BED 1           530.0      MIN T FOR RATE EQUN   345.0
   MIN WORKING T BED I (I<N)   400.0      T.R.F.               1699.80
   MAX WORKING T BED I (I<N)   530.0
   MIN WORKING T BED N         400.0
   MAX WORKING T BED N         530.0

DATA FOR 2 BED NH3 COLD SHOT CONVERTER
   INLET PRESSURE (ATA)        140.0        -           INLET
   EXIT NH3 (%)                 12.001      -           H2(%)   63.330
   MAKE (TE/DAY)               868.00       -           N2      20.994
   TOTAL FLOWRATE (RM3/HR)  576441.          -          CH4     10.847
   MIN OVERALL DT (DEG.C)       60.0        -           A        2.753
   T COLD SHOT (DEG.C)         142.0        -           HE       0.000
   CATALYST ACTIVITY  67.2 % NEW. BULK DENSITY 2.63 TE/M3  NH3   2.076
   CATALYST SIZE (MM)    6.0

RESULTS
                H2(%G)    N2      CH4      A       HE      NH3     T(DEG.C)
BED 1
   INLET        63.330  20.994  10.847   2.753   0.000   2.076     400.0
   EXIT         57.603  19.077  11.564   2.935   0.000   8.822     505.4

        CAT VOLUME (M3)        22.466     T EQUILIB (DEG.C)   522.5
        G INLET (RM3/HR)    409289.1      SHOT (% G TOTAL)      0.00
            EXIT            383918.5             (RM3/HR)        0.0

BED 2
   INLET        59.340  19.659  11.346   2.880   0.000   6.775     400.0
   EXIT         54.903  18.174  11.902   3.021   0.000  12.001     475.8

        CAT VOLUME (M3)        42.936     T EQUILIB (DEG.C)   485.0
        G INLET (RM3/HR)    551070.2      SHOT (% G TOTAL)     29.00
            EXIT            525359.8             (RM3/HR)   167151.8


TOTAL CATALYST VOLUME (M3)      65.40    EXIT NH3 (%)          12.001
TOTAL FLOWRATE (RM3/HR)      576441.0    MAKE (TE.NH3/DAY)    868.00
TOTAL COLD SHOT (%)              29.00
```

FIG. 74. Computer form—NH_3 cold-shot converters/design.

output are in metric/Centigrade units, and up to eight catalyst beds may be specified. Though the program is ostensibly only for cold-shot converters, it may be used also for design of indirectly cooled converters, by specifying a fictitious ($< -273.15°C$) cold-shot temperature. An interesting effect of the ability to specify a minimum overall temperature rise constraint is that in some instances the program will indicate (correctly) that the optimal design is one having fewer beds than specified. In other words, it will give certain bed volumes and shot rates as zero.

RTK26 program

The ammonia cold-shot converter optimal performance program is designated RTK26. This broadly parallels RTK25, though it is not possible with RTK26 to require a minimum overall temperature rise. As for the shift converter performance program, the solution will, wherever possible, be within the working temperature constraints, and where this is not possible, bed inlet temperatures will be set to the minimum permissible temperatures. A specimen of the data form and output from the program are shown in fig. 75.

Running times are, for RTK26, about one minute per bed on an IBM 360/50 computer. For RTK25, times are about the same where an overall temperature rise constraint is effective, and about half as much when such a constraint does not operate.

RTK22 program

The ammonia tube-cooled converter program is designated RTK22. It is intended for calculations on counter-current (TVA type) converters and, as mentioned earlier, may be used for either design or performance calculations. Essentially, it is a straight-through performance program with an external optimising routine. The input data comprise the inlet gas rate and composition, the synthesis pressure, the rate of direct bypass (if any) or the heat input to the synthesis gas, the bed inlet temperature, and the 'bed cooling factor' (which is the area of cooling-tube surface per unit catalyst volume multiplied by an appropriate heat transfer coefficient). The data must also include a terminal condition—either a catalyst volume or an exit concentration (which may be expressed as te NH_3/day). Since all conditions at bed inlet are defined, integration through the bed of the reaction kinetic equation, the heat-balance equation, and the heat-transfer equation, may be effected without regard to the bed exit (terminal) condition. It is for this reason that alternative terminal conditions, and the use of the program for either performance or design calculations, is possible.

Where bed inlet temperature and bed cooling factor are specified, the program will execute only a single calculation. If, however, either is not specified, and is entered as '−1' in the input data, the program will search for an optimum. Thus it will optimise with respect to bed inlet temperature, to bed cooling factor, or to both. By taking as the quantity to be optimised the ratio (exit concentration/bed volume), the optimisation will be either of design or of performance, depending on which of the two criteria is fixed and used as a terminal condition.

Automatic checks ensure that percentages sum to 100, that the problem is not overspecified (both bed volume and exit concentration being given, for

COMPUTER PROGRAMS FOR CONVERTER CALCULATIONS

OPTIMAL PERFORMANCE

		Program	RTK 26
		O.S.	4
			P
		Time (mins)	
		Job No.	

		Units	Computer Data			
Title			TEST 1 (ARTK26)			
Date						
Th.Exp.Ref.No.						
20	Number of copies		2.			
21	Number of beds		4.			
22	Inlet :	H_2	%	65.349		
23		N_2	%	23.178		
24		CH_4	%	7.071		
25		A	%	1.829		
26		He	%	0.		
27		NH_3	%	2.573		
28	Pressure		ata	116.2		
30	Inlet gas rate		Nm³/hour	463000.		
32	Cold shot temperature		°C	140.		
33	Catalyst size		mm.	6.		
34	Catalyst age		years	0.6		
35	Impurities in M.U.G. (eg. O_2 Cpds.)		p.p.m.	5.		
36	Catalyst volume ;	bed 1	M³	9.5		
37		2	M³	12.55		
38		3	M³	18.25		
39		4	M³	24.7		
40		5	M³			
41		6	M³			
	Control number			-1.		

ADDITIONAL RELEVANT INFORMATION:

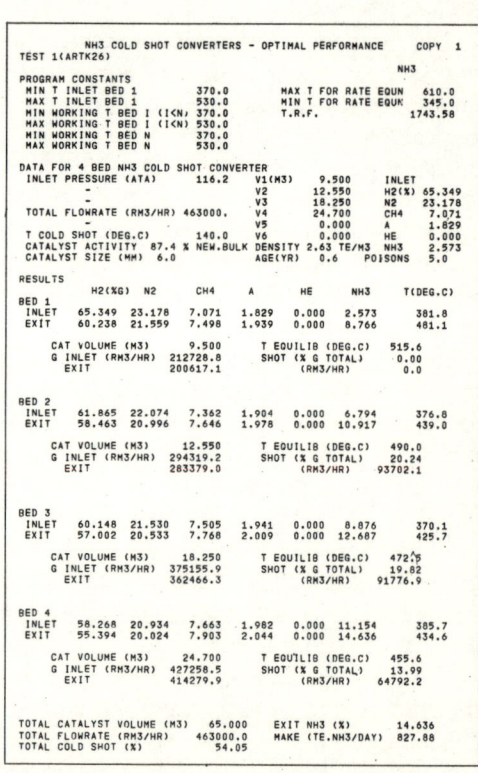

FIG. 75. Computer form—NH_3 cold-shot converters/performance.

COMPUTER PROGRAMS FOR CONVERTER CALCULATIONS

		Units	Computer Data				
					PROGRAM		RTK22
					OS		4
							P
					TIME (mins.)		
					JOB No.		
Title			TEST 3 (ARTK22)				
Date							
Th.Exp.Ref.No.							
Ammonia Make		Te/Day	-9.				
Pressure inlet the converter		atas	250.				
Pressure drop across converter		atas	3.				
Catalyst size		mm	8.				
Catalyst age		years	100.				
Impurities in M.U.G. (e.g. O_2 Cpds.)		p.p.m.	-9.				
Flow rate inlet the converter		$\bar{R}\,M^3/hr.$	50000.				
Bypass temperature		°C	35.				
Inlet:	H_2	%	64.				
	N_2	%	20.				
	CH_4	%	8.				
	A	%	3.				
	He	%	0.				
	NH_3	%	5.				
Bypass flowrate		$\bar{R}\,M^3/hr.$	0.				
Bed inlet temperature	**	°C	460.				
Catalyst volume	*	M^3	3.6				
Exit NH_3	*	%	-9.				
Bed Cooling factor	**		20300.				
Control symbol			-1				

ADDITIONAL RELEVANT INFORMATION

```
    TUBE COOLED AMMONIA CONVERTER      COPY    1
       TEST 3(ARTK22)
                               NH3/

MAKE                88.6 TE/DAY
INLET PRESSURE     250.0 ATA       PRESSURE DROP      3.0 ATA
INLET GAS RATE   50000. RM3/HR     BYPASS GAS RATE    0.0 RM3/HR
TEMPERATURE         35.0 DEG. C    T.R.F.          1697.68 DEG. C

                   H2       N2      CH4      A       HE      NH3
INLET ANALYSIS  .64.000  20.000   8.000   3.000   0.000   5.000 %
EXIT ANALYSIS   53.993  16.509   8.931   3.349   0.000  17.217 %


BED INLET TEMP       460.00 DEG. C
CATALYST VOLUME        3.60 M3
COOLING FACTOR     20300.00
CATALYST ACTIVITY    100.00 % NEW.BULK DENSITY 2.80 TE/M3
CATALYST SIZE          8.00 MM

TUBE TEMP   CAT TEMP   NH3 CONC   CAT VOL
  460.        460.        5.00      0.00
  456.        509.        8.41      0.36
  446.        532.       10.61      0.72
  433.        537.       11.84      1.08

  418.        533.       12.64      1.44
  401.        526.       13.32      1.80
  383.        517.       14.01      2.16
  363.        507.       14.74      2.52
  343.        497.       15.51      2.88
  320.        486.       16.34      3.24
  297.        474.       17.22      3.60
```

FIG. 76. Computer form—tube cooled ammonia converter.

example) and that physically impossible calculations are not run. Facilities exist within the program for taking account of heat transfer not only to the cooling tubes, but also to the sheath gas. This last is, however, customarily ignored—the amount of heat transferred is very small, since the catalyst bed is lagged, and the representation of the bed cooling by a rolled-up 'bed cooling factor' obviates the need to specify a particular geometry at the design stage. Given the bed volume which the program produces, the converter geometry is settled later on the basis of the allowable pressure drop.

A specimen data form and computer output are shown in fig. 76. The program is written to accept data and to output results in metric/Centigrade units, and a typical running time is of the order of one minute on an IBM 360/50 computer.

Other programs

A number of other computer programs exist which are related in one way or another to the converter programs listed above. There are, for example, programs for the calculation of mass and heat balances for steam reforming systems, which can be used to define the input streams to subsequent shift conversion systems. There are, similarly, programs for mass and heat balances in ammonia synthesis loops, which will indicate design requirements for the associated synthesis converter. Indeed, there is a program (RU009) which combines synthesis loop mass and heat balances with the converter design, to facilitate process design studies. Part of the output from this program is a diagrammatic representation of gas flows and temperatures around the synthesis converter.

In a different direction, programs exist for the calculation of pressure drops through catalyst beds, for the physical sizing of ammonia synthesis converters for minimum vessel cost, and so on. That is to say, programs concerned with the mechanical engineering aspects of converter design. Since these necessarily incorporate local practice, they may be of less general interest. Their utility for problems of converter design is, however, unquestionable.

An alternative expression of the techniques which have been developed is in the construction of programs for the simulation of operating converters, whether to assess catalyst activity and performance over life, or to pin-point trouble spots—for example, mechanical damage to converter internals. In a number of instances, a variant of the Simplex technique has been used for the least squares fitting of the often inconsistent data obtained from plant records. The same least-squares technique is used (Porter and Snowdon, in press) for the analysis and fitting of laboratory data on the kinetics of new or improved catalyst formulations—a continuing exercise which is eventually reflected in the 'front-line' programs described in earlier sections, and in similar programs for other reactions. At all stages, therefore, from the laboratory through to the routine handling of customer enquiries, the treatment of catalyst problems increasingly reflects and relies on the facilities of the digital computer and the new techniques developed to take advantage of these. This is perhaps the salient feature of the current technology.

CHAPTER 9

Handling and using catalysts on the plant

IN a book of this type it may seem unnecessary to say that the efficient operation of a chemical plant depends to a considerable extent on the condition of the catalyst employed. For example, the gas flow through catalysts described in Chapter 3, and hence their performance, depends on correct packing. Similarly, catalyst performance and lifetime are related to careful reduction and operation. The success of the process is closely linked to the way in which catalysts are handled. Unfortunately, the practical preoccupations occupying the minds of people involved in the construction, commissioning, and operation of new plants are mainly of a mechanical nature, and proper treatment of the catalyst does not always receive the attention it deserves. Catalyst handling should be considered right from the design and construction planning stages, so that suitable manholes, access platforms, lifting beams, chutes, sieves, hoppers, and so on, are provided to make it possible for catalysts to get the treatment their importance warrants.

Catalyst charging
Drum handling

Catalysts are generally supplied in drums, the full drums weighing between 80 and 250 kg. Metal drums are usually suitable for outside storage for a few months, but should be protected against rain and puddles. If prolonged storage is expected, or if the drums are not metal, they should be kept under cover, and away from damp walls and floors. The lids should be left on the drums until just before the catalyst is charged, and if lids are accidentally knocked off it is important that they be replaced as soon as possible, for contamination of the catalyst must be avoided.

Catalysts should be handled as gently as possible. Suitable space is required for storing drums between delivery and charging, and double handling can be avoided if this space is close to the reactors. When a crane is used for stacking or destacking drums, or for lifting them to the charging manholes, suitable standing must be available. If a small mobile crane, or a forklift truck, is used, a smooth paved area is desirable. When drums are lifted to the charging point it is generally most satisfactory to use a mobile crane rather than an individual lifting beam on the vessel, because a crane can lift drums off a wide area, and this avoids multiple handling. Drums should not be rolled, so if manhandling is unavoidable, suitable drum barrows, upending levers, and skids, should be provided. More importantly, proper equipment is essential for the safety of

FIG. 77. Drum-handling equipment. (Courtesy of Powell & Co.)

the workers—particularly when the larger drums are used. Some typical equipment is illustrated in fig. 77.

Sieving

Catalysts are sieved before they are packed into their drums for despatch, but some attrition may occur in transit if the drums are roughly handled, so some form of screening is usual before charging, especially if, on delivery, the

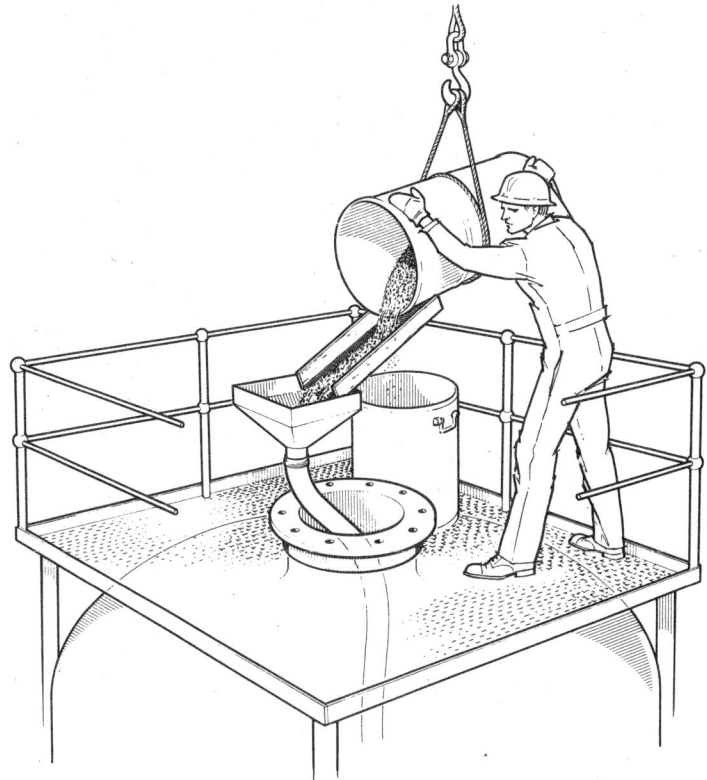

FIG. 78. Mechanism for transferring catalyst to charging hopper.

catalyst appears to contain dust. A simple method of screening is to pass the catalyst over an inclined chute made of a suitably sized mesh, and this is often the most satisfactory method, since vibrating screens can cause further unnecessary fragmentation and loss of catalyst. A suitable arrangement, which also decreases handling, is shown in fig. 78. This system employs a sieve to convey the catalyst from the drum to the charging hopper or chute. Another useful technique is to winnow the dust away with a compressed air jet while the catalyst is pouring into the charging bucket or hopper. Simple hand-sieving may be used, but the operators must be instructed to stop sieving when the dust has been removed, so that unnecessary attrition is avoided.

Charging vessels

Two general rules are usually recommended for charging catalyst into vessels. The catalyst should not have a free fall of more than 60–100 cm (2 to 3 ft), and it must be distributed evenly as the bed is filled. The distance that a catalyst can fall without serious damage depends on its strength and shape. A hard, spherical pill withstands dropping better than a soft, angular pellet or chip. The catalyst should not all be poured into the vessel in one spot and the resultant heap raked level. If this is done the particles tend to segregate. Small particles and dust mainly stay in the centre of the heap, whilst the larger pieces roll to the edge. Such segregation leads to uneven gas distribution.

Uneven packing and breakage during charging seriously affect the gas distribution and the utilisation of the catalyst. The degree of packing in the catalyst bed has a marked effect on the voidage, so that in a bed of regular pellets the voidage can vary by over plus or minus 10 per cent, according to the degree of packing. If the granules are not all the same size, or if fines are present, the voidage variations can be considerably greater. The effect of packing on pressure drop is even more marked, because the pressure drop is roughly proportional to $(\text{voidage})^{-3}$. Thus the difference in the pressure drops between loosely packed and closely packed beds of regular pellets is roughly in the ratio $90^3 : 110^3$, or about 1 to 2. In practice the pressure drop increases by about 50 per cent when the catalyst bed settles during use.

The effect of packing variations (voidage) on the distribution of gas flow is yet more pronounced than its effect on pressure drop. An extreme example would be that of two identical reactors in parallel which are packed to the same depth, but to different voidages—that of the one being 20 per cent higher than that of the other. In such a system, the ratio of the gas flows is proportional to the sixth power of the ratio of the voidages. This, in fact, means that 77 per cent of the gas passes through one converter; so the space velocity has become 154 per cent of the intended value. The remaining 23 per cent of the gas flows through the parallel converter at only 46 per cent of the intended space velocity. In practice, maldistribution of packing sometimes occurs when all the catalyst is allowed to fall into one part of the bed, just beneath the charging manhole, or if the operators walk on the bed, or rake it during the charging process.

A deep, narrow bed is the easiest shape to charge satisfactorily, and irregularities are more likely to average out along the direction of flow. Extra care is required with wide beds because it is difficult to distribute the catalyst evenly over a large area. Radial flow beds are particularly susceptible to irregular packing and uneven gas flow.

Another danger of irregular packing and uneven gas flow can arise in reactors which depend on the flow of gas to control the temperature of the bed. Examples include the hydrogenation of acetylene in the presence of ethylene, also some types of shift converter, where the reaction is not expected to reach equilibrium. High local temperatures will occur in regions of low flow if the reaction proceeds further than desired.

Before catalyst is put into a vessel it is important that the condition of the catalyst support grid should be checked, because faults at this point are difficult to rectify after charging. Some form of light metal shield, or 'spider', fitted into the discharge manhole, prevents the branch from forming a stagnant space full of catalyst.

One of the quickest ways of filling a vessel evenly is with a canvas sock fitted to a hopper which is supported outside the manhole. The sock is kept full of catalyst and raised slowly to allow the catalyst to flow into the vessel in a controlled manner. The sock must be guided in some way so that it does not always discharge at the same point, and, to avoid kinking, it usually needs to be shortened periodically as the vessel is filled. A metal tube, which has the advantage of being more easily steered, can also be used, and is better for charging the heavier and more abrasive agents, including the ammonia synthesis catalyst. The tube is usually made in flanged sections, which can be removed as the vessel is filled and a shorter tube is needed. Another successful refinement has been a hinged flap fitted at the bottom of the charging tube, which can be operated with a cord from the top. The catalyst flow can then be readily controlled.

When operators have to enter the vessel during or after charging, planks should be used, so that they do not tread directly on the catalyst. Most catalysts produce dust in the vessel during the charging process, and dust masks must be provided for anybody who has to enter at this time. Protection against the dust is particularly important if the material has any toxic properties, as in the case of high temperature shift catalysts containing chromia.

If the catalyst bed is wide and access is through a side manhole, it can be difficult to get the catalyst distributed across the vessel without raking. But, as mentioned above, raking is particularly undesirable as it leads to a concentration of fines near the charging manhole. A small conveyor, which is narrow enough to fit through the manhole and long enough to reach across the vessel, is useful under such conditions. Suitable portable conveyors are readily available.

A vessel can also be charged by the rope and bucket method. This may involve having a man standing on a plank inside the vessel to empty the bucket on to the bed. Alternatively, a double-rope system may be employed, so that the bucket can be tipped, at the appropriate moment, from outside. There are also buckets with opening bottoms, and a successful design is shown in fig. 79. Sometimes the catalyst drums themselves can be lowered and emptied into the vessel. This cuts out one handling stage, but denies the user the chance of screening the catalyst before charging. It also involves the risk of damage to the inside of the vessel from the heavy drums—a particular risk in vessels having a refractory lining. If buckets are used, they should be made of rubber or plastic rather than metal. And it is, of course, important that their handles are securely fixed.

It is often useful to put a holding-down grid, or layer of heavy inert material, on the top of the catalyst bed to prevent movement of the pellets by high velocity gases. An inert layer of ceramic rings can be useful in intercepting any contaminants in the gas stream before they reach the catalyst itself. In the case of secondary reformers, the inert material must be silica-free to avoid silica deposition in the subsequent plant.

When the catalyst is charged into tubes, it is again important that the catalyst is not dropped more than about 40 cm (a foot or so). The catalyst pellets must not be allowed to form bridges across the tubes, as this leads to gaps in the catalyst, and overheating of the tubes during operation. Before charging begins, the inlet and exit connexions must be checked to ensure that they are clear. The interior of the tube must be smooth. Irregularities could provide support for catalyst bridging that might build up during operation.

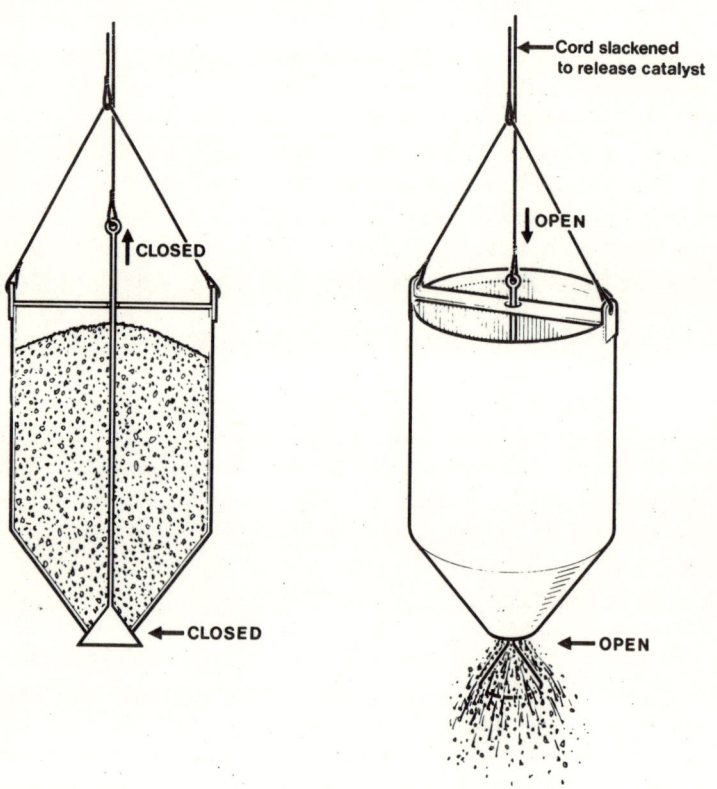

FIG. 79. Charging bucket.

A useful check for obstructions in the inlet and exit connections can be made with the air-flow device shown in fig. 80. The expanding bung is clamped into the tube and the air supply opened out to maintain a certain pressure (say 30 psig) at the upper side of the orifice plate. The pressure on the downstream side of the orifice then provides an indication of any obstruction in the gas exit connection. The same device is used to check the pressure drop through the catalyst after charging.

The method used to fill the tubes depends on the access available. Tubes that have a bolted flange at the bottom as well as the top can be filled by the 'sweep's-brush' method. A sectioned rod is attached to the bottom catalyst support, which is then pushed up to the top of the catalyst tube, and free movement of the catalyst support is checked at the same time. Additional rods are added as the catalyst support is moved up the tube. Bamboo brush-rods or drain-rods with screwed ends are suitable for this process. When the support grid is almost at the top of the tube, catalyst is poured in and the grid lowered slowly by withdrawing the rods to maintain the catalyst surface a foot or so below the top flange.

The catalyst should be weighed before charging to provide a check on the

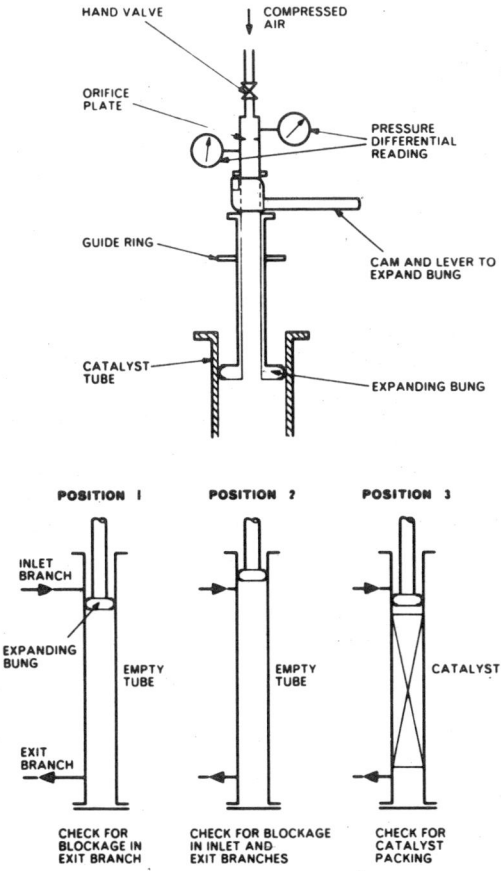

FIG. 80. Air-flow device for checking pressure drops through catalyst.

quantity charged to each tube. The tubes should be vibrated periodically during the charging process, either with electric vibrators fastened inside the furnace or by hammering the top flange with hide-faced hammers. The final catalyst level is measured carefully to ensure that all the heated length of the tube will contain catalyst when the plant is at operating temperature.

When charging has been completed, the pressure drop and charged-weight figures should be examined. All pressure drops should be within about 5 per cent of the mean. A higher figure indicates broken rings, in which case the whole tube should be discharged and recharged after the catalyst has been sieved. A low pressure drop suggests that there are gaps in the catalyst, and that the tube requires further vibration. Pressure drop variations are usually reflected in the charged weight, but this is not such a good indication of poor packing as the pressure drop.

Reaction tubes which are closed at the bottom are best packed with catalyst by the 'sock' method. The catalyst is first put in a fabric tube, or sock, about 150–200 cm (5–6 ft) long, sized so that the filled sock slides easily inside the reaction tube. A strong cord is tied round the top end of the sock and about 9 in at the bottom end is folded up to hold the catalyst in place whilst the sock is lowered into the tube. When the sock has reached the bottom of the tube, the cord is withdrawn gently, which allows the bottom end of the sock to unfold and the catalyst to fall into place. It is important that the cord should be longer than the catalyst tube, and it is useful to anchor the top end of the cord so that it cannot fall into the tube. After the tube has been charged it is topped up, and the pressure drop is checked, just as with the 'sweep's-brush' method. Some type of extraction gear should be available before charging is begun in case some of the tubes have to be discharged to correct high pressure drops, or to remove socks which occasionally become stuck inside the tubes. The socks can be filled at ground level and lifted to the furnace top in a sling, which both avoids the need to lift the drums themselves on to the furnace top, and also generally simplifies handling.

ICI ammonia synthesis catalyst is a particularly robust material, and can be tipped directly into most types of tube-cooled converters without suffering significant damage. The top ends of the cooling tubes should be plugged or covered in rows with a small channel section to prevent the catalyst from blocking them during the charging. The tube spacer plates prevent excessive free fall, but also hinder the introduction of charging tubes. The catalyst should be tipped evenly round the top of the cartridge to prevent heaping and segregation in the bed. Some forms of pre-reduced synthesis catalyst are more fragile than the standard type, and are charged with a tube and hopper. A tube and hopper is also required for charging cold-shot converters. The tubes should be of metal because of the dense and abrasive nature of the catalyst.

Most types of synthesis catalyst consist of irregular chips, and the bulk-density can vary with the method of packing. The bulk-density should be checked from time to time as the converter is filled, to ensure that the correct weight and volume will be achieved for the charge. If the density is too low, the best remedy is to drop the catalyst from a greater height on to the bed. The falling catalyst then packs more closely, and without significant attrition so long as the drop does not exceed about 4–5 m (15 ft). A low bulk-density, maintained throughout charging, would mean that a low weight of active material was charged, which might result in poor performance of the converter and a short life for the charge.

Catalyst reduction

Catalysts are designed to be in their active form under process conditions. Generally they are supplied in an inactive form which is stable under atmospheric conditions, and sometimes they also contain small quantities of impurities, such as sulphur, which have to be removed in the course of activation.

The activation procedure, usually involving reduction of an oxide, requires careful control if the best results are to be achieved.

Reforming catalyst reduction

Nickel reforming catalysts are supplied as supported nickel oxide, and to become active they must be reduced to give metallic nickel. In some cases they contain traces of residual sulphates, which must be removed, by reduction with hydrogen in the presence of steam before full activity can be achieved. In a typical reduction schedule the catalyst is heated by circulation of a dry gas until the temperature is above that for steam condensation. Steam is then added,

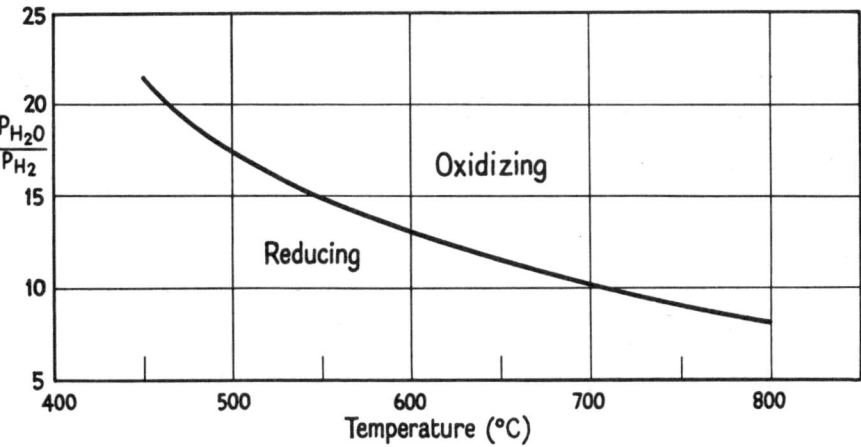

FIG. 81. Reducing conditions for nickel oxide.

after which the heating is continued until the operating temperature is reached. During this time hydrogen is introduced to maintain a ratio of steam to hydrogen below 7:1. The steam/hydrogen ratio required for reducing nickel oxide varies with temperature, as is shown in fig. 81. The nickel is reduced with a lower proportion of hydrogen at lower temperatures, but reduction is generally continued right up to the operating temperature to ensure that the sulphur level is at a minimum. The steam appears to assist in sulphur removal, as well as lessening the rise in temperature by diluting the hydrogen, and only a little sulphur will be removed until both hydrogen and steam are present and the temperature is above 700°C. In an ammonia plant the secondary reformer catalyst is reduced without difficulty at the same time as the primary catalyst.

Natural gas reformers are commonly brought on line with natural gas itself, relying on the thermal cracking of methane to provide the hydrogen for reduction. The gas is introduced at a low rate at about twice the normal steam ratio, and the gas rate is increased slowly, with the steam held constant, until the exit methane concentration shows that the catalyst has become active. Natural gas can also be used during the initial heating of the reformer before steam is introduced so long as the temperature does not exceed 300°C. In the absence of steam, there is a risk of carbon laydown at higher temperatures.

High-temperature shift catalyst reduction

High-temperature shift catalysts are supplied in the form of Fe_2O_3, containing traces of sulphur in the form of sulphate. During reduction, the Fe_2O_3 is converted to Fe_3O_4, and the sulphur is removed as H_2S.

In plants where the shift converter follows a reformer, the catalyst is usually reduced in the course of the reformer start-up. Reduction of the shift catalyst can begin at about 150°C, whilst the reforming catalyst is being reduced with hydrogen and steam, and in most cases reduction will be continued after the hydrocarbon feed is added. In an ammonia plant the shift catalyst often does not reach full operating temperatures until air is put on the secondary reformer. Natural gas plants which omit the steam and hydrogen stage are able to start to reduce the high temperature shift catalyst as soon as hydrogen appears in the reformer effluent, so it is most important to ensure that steam is also present when the high temperature shift catalyst is being reduced. High temperature shift catalysts must not be exposed to hydrogen in the absence of steam, even when they have been reduced to Fe_3O_4, as further reduction to iron is possible (see pp. 101 and 102).

When the high temperature shift converter follows a new charge of primary catalyst containing traces of sulphur, it is preferable that the gas leaving the primary (when this is possible) is not put into the high temperature shift until the sulphur content is less than 1 ppm. No harm will be done to ICI shift catalyst, which will work very well in a sulphided state, but the shift catalyst will retain the sulphur for a time and delay the elimination of sulphur from the system (see p. 103).

The reduction of Fe_2O_3 to Fe_3O_4 with hydrogen is only slightly exothermic and no significant temperature rise is likely unless CO is also present. The reaction

$$CO + H_2O \rightarrow CO_2 + H_2 \quad \Delta H = -9.83 \text{ kcal}$$

will increase the temperature of the gas by about 10°C for each 1 per cent of CO present in the wet gas. Above 630°C there is a phase change in the Fe_3O_4 which rapidly leads to a permanent loss of activity; some loss of activity can occur if the catalyst is heated above 540°C.

High-temperature shift converters are used in many other types of plant, and a typical reduction schedule is as given below for use with catalysts 15–4 and 15–5, or other catalysts that can withstand condensing steam.

(1) Purge the converter free from air using an inert gas.
(2) Start the flow of process steam and gas to give a space velocity of about 200 h^{-1}. The process gas stream should contain less than 15 per cent carbon monoxide (in the wet gas), and this can be controlled by adjusting the steam rate. The converter usually requires periodic draining to avoid accumulation of condensate. Heat the converter slowly at about 50 to 60°C per hour until the inlet temperature reaches 300°C. Reduction usually commences at about 150°C and is normally almost complete at 300°C.

(3) When the first part of the bed reaches 300°C, reduce the amount of heat supplied to the converter, since temperatures will rise when the exothermic CO shift reaction begins.
(4) Temperatures must be kept below 500°C during reduction by the addition of extra steam if required. The shift converter should be brought to operating conditions slowly to ensure that it is working at the lowest possible temperature consistent with a satisfactory exit composition.

Low-temperature shift catalyst reduction

Low-temperature shift catalysts, supplied as supported copper oxides, require careful reduction to give the maximum activity. Sintering, with a consequent loss of surface area and activity, occurs rapidly at temperatures above 280°C, and 250°C is normally regarded as the maximum operating temperature. Converter temperatures should be kept below 230°C during the exothermic reduction to prevent damage from sintering, because local overheating of the catalyst surface may occur (without significant rise in any measured temperature) and reduce its activity. The catalyst is reduced by being slowly heated in a controlled, low concentration of hydrogen. The following reduction schedule for ICI catalyst 52–1 is typical.

(1) Purge the catalyst free from air with an inert gas.
(2) Heat the catalyst in the inert gas stream to 120°C at about 50°C/h.
(3) Pass the inert gas through the catalyst with a space velocity of about 600 h^{-1} and add hydrogen gradually up to 0·5 mole per cent. A circulating system is often employed to economise in the use of gas at this stage, but, when natural gas is used as the inert carrier, it is normally blown off after leaving the reactor.
(4) Increase the bed temperatures at about 30°C/h to 180°C.
(5) When reduction begins, the temperature at the inlet will rise, and a hot front will travel down the bed. After conditions have settled, raise the temperature to about 200°C in the course of one hour.
(6) Slowly increase the hydrogen concentration to 1·5 per cent over a period of about 3 hours. Whenever the hydrogen concentration is increased, the catalyst temperature will rise again, and a hot front will again proceed down the bed. The maximum temperature must not exceed 230°C. It may be necessary to reduce the inlet temperature to assist in temperature control. Towards the end of the reduction the temperature rise across the converter will fall and the inlet temperature should be raised to maintain the maximum temperature at about 225°C. It is valuable to check the quantity of hydrogen consumed in the course of reduction. For ICI catalyst 52–1 about 80 m^3 of hydrogen are required per m^3 of catalyst. If for any reason it seems that the expected quantity of hydrogen has not been used, the hydrogen level should be gradually increased until complete reduction is achieved, still keeping the maximum catalyst temperature below 230°C. The reduction water can be collected, as its measurement provides another check on the progress of reduction. About 60 kg of reduction water should be evolved per m^3 of catalyst and there is likely to be a similar quantity of absorbed water.
(7) Finally, raise the inlet temperature to 225°C and allow the hydrogen

concentration to increase slowly to 10 per cent or more, if this is possible, to ensure complete reduction. Very little water will be formed at this stage, and no temperature rise should be observed. Catalyst temperature must still be kept below 230°C.

It is not essential to employ a space velocity as high as 600 h^{-1} during reduction, but this minimises the time involved, and temperature control is simplified at gas rates of this order.

Methanation catalyst reduction

Methanation catalysts are activated with hydrogen, reducing nickel oxides to metallic nickel. The vessel is first purged free of oxygen with nitrogen or process gas, then heated at about 20–50°C per hour to 200°C. The initial heating is sometimes done with an inert gas such as nitrogen or natural gas, but frequently process gas is employed. Process gas can be used when the catalyst is unreduced without any danger of nickel carbonyl formation (see section on safety precautions). Reduction commences at about 200°C, and the reducing gas is usually process gas. The total $CO + CO_2$ content of the reducing gas must not exceed one per cent, and the composition of the inlet gas must be controlled at or below this figure by careful attention to the preceding plant. The methanation reaction will commence as soon as some of the catalyst is reduced, and the catalyst temperature will kick up by about 6°C for each 0·1 per cent CO_2 and 7·4°C for each 0·1 per cent CO. It is best to raise the inlet temperature at about 25°C per hour to 350°C, so that the catalyst is not overheated. The maximum operating temperature of methanation catalysts is about 425°C, but they are not damaged by brief overheating. The vessel design, however, will generally place some limit on the maximum allowable temperature. The best way to cool the catalyst and to even out hot spots is to increase the flow of process gas, provided that the $CO_2 + CO$ content is less than one per cent. It is useful to increase the catalyst temperature to 400°C at the end of reduction to ensure that maximum activity is reached (catalyst 11–3 has good activity before this treatment). If it is not possible to reach 400°C by adjustment of the temperature controls, catalyst temperatures can be increased by by-passing some of the gas flow round low-temperature shift converter, so as to increase the CO content of the gas entering the methanator and thus the temperature rise across it.

Ammonia synthesis catalyst reduction

Ammonia synthesis catalysts are usually supplied in the form of Fe_3O_4, which must be reduced to metallic iron before it is catalytically active. Some catalysts are supplied in a pre-reduced form, with over 90 per cent of the catalyst already reduced. They require a much shorter time for reduction, but the same general considerations apply. Normal catalyst is reduced with hydrogen or synthesis gas by the following reaction

$$Fe_3O_4 + 4H_2 \rightarrow 3Fe + 4H_2O \quad \Delta H = 35.81 \text{ kcal.}$$

External heat must be supplied if the endothermic reaction is to be done with hydrogen alone. But when reduction is carried out with synthesis gas, ammonia is produced as soon as the first of the catalyst is reduced, and this decreases the external heat requirements. The ammonia synthesis reaction is exothermic and provides heat to raise the catalyst temperature and accelerate the rate of reduction. As soon as the catalyst temperatures show signs of rising on their own accord, the gas rate through the bed can be increased to transfer the heat down to the bed and assist in reducing the remainder of the catalyst. As more ammonia is formed the bed becomes self heating, and the heat input can be decreased until eventually, when all the available gas is going through the catalyst, a cold by-pass will be required in order to control the inlet temperature, and prevent an excessive rise in the temperature of the hot spot.

Catalyst activity is lost by operation at temperatures above 550°C, and it is desirable to keep temperatures below 500°C during the reduction. The water which is produced during reduction can deactivate part of the catalyst which has already been reduced, and this effect is more severe at temperatures above 450°C (Chapter 7).

It is normal to measure the water content of the gas leaving the catalyst, and to keep this water content below 20 000 ppm by adjusting the rate of temperature rise in the converter. The permissible water content is related to the converter design, and the available gas rate, and under some conditions it is necessary to hold the water level below 5000 ppm. The circulating gas leaving the ammonia converter must be cooled in order to condense out as much of the water as possible before the gas is put back over the catalyst. High-pressure operation assists in the condensation of water, and the ammonia produced when using synthesis gas ensures that the condensate is a water/ammonia solution, which again assists in the removal of water. If the loop contains a refrigeration cooler this can, with advantage, be commissioned as soon as the ammonia content of the condensate is sufficient to avoid freezing. The ammonia content of the reduction water usually rises rapidly in the first few hours, and will exceed 20 per cent before much reduction has taken place. The freezing point of 20 per cent ammonia in water is $-33°C$ which is the lowest temperature likely to be met in a refrigerated loop. If the reduction schedule is interrupted, it is very important to maintain a forward flow of gas so that water cannot diffuse back on to the freshly reduced catalyst. This is normally ensured by closing a valve on the converter inlet line and blowing off gas from the exit line.

A typical reduction procedure for ICI 35–4 in a tube cooled ammonia converter is given below.

(1) Bring the loop to full operation pressure.
(2) Start gas circulation round the loop and adjust flow through heater and catalyst.
(3) Commence heating this catalyst at about 50°C/h to 340°C. The gas rate should be as high as possible to assist in the heating process. Some reduction may start at about 280°C.
(4) Decrease the heating rate to 10°C/h, preferably by an increase in the gas flow rate rather than a cut in the heater output.
(5) Measure the water content of the gas leaving the converter and control it within the range 5000–10 000 ppm by adjusting the rate of increase in temperature.
(6) Keep the peak catalyst temperatures below 500°C until the reduction is virtually complete.

Reduction of a multi-bed quench converter follows somewhat different lines and a typical procedure is as follows:
(1) Bring the loop to full operation pressure.
(2) Start gas circulation round the loop and adjust flow through heater and catalyst.
(3) Heat the whole converter at 50°C/h until reduction water is first observed, generally when the highest temperature reaches about 300°C.
(4) Continue heating the converter to 340°C at 50°C/h, but hold the second and later beds at about 300°C, or at whatever temperature reduction is first observed.
(5) When the top bed temperature reaches 340°C decrease the rate of heating to about 10°C/h.
(6) Check the exit water concentration and control in the range 5000–10 000 ppm by adjusting the rate of heating.
(7) When water evolution from the top bed falls off, the second bed can be heated up in the same way as the first, keeping the third and later beds at about 300°C.
(8) Reduce the remaining beds in turn, as for the second.

Regeneration of catalysts and blanketing procedures

High-temperature shift catalyst regeneration

High-temperature shift catalysts may become contaminated with gums and carbon formed in the preceding plant, and may be regenerated by treatment with steam and air.

Superheated steam should be passed over the catalyst at 400–500°C and at a space velocity of up to 1000 h^{-1}. The steaming is continued until the concentration of hydrogen sulphide has been reduced to less than 80 or 100 ppm by volume. The inlet temperature is then dropped to 300–350°C, and air is admitted cautiously to give not more than one per cent of oxygen in the steam. It is important that no temperature should exceed 550°C, and if temperatures are rising and appear likely to exceed 500°C, the air supply should be cut off, and the catalyst allowed to cool in the steam flow. Progress during regeneration can be followed by measuring the CO_2 content of the exit gas. Some of the catalyst will be oxidised at the same time as the carbon compounds are being removed, and will require re-reduction before it can be used again.

The catalyst cannot be regarded as completely oxidised when the carbon has been removed.

High-temperature shift catalysts may also be contaminated with water-soluble deposits from the preceding plant. With the more dense catalysts such as ICI 15–4 and 15–5 these deposits may be removed by washing with water without imparing the catalyst activity. The water is best passed through in the opposite direction to that of the normal gas flow, so that the contaminants are washed out of the bed rather than into it. The catalyst is normally cooled to about 50°C before the introduction of water.

Reforming catalyst regeneration

The operation of reforming catalysts is sometimes impaired by carbon deposition, usually as a consequence of mal-operation. These deposits may be removed either with steam alone, in the absence of feedstock, or with a steam/air mixture. The normal procedure is to take off the hydrocarbon feed and adjust the firing to keep the temperatures at about the normal operating level. The steaming can be continued for several days if convenient, but about six to 12 hours is usually sufficient. Some re-reduction will be necessary after a prolonged period of steaming, but, after a short regeneration, it is usually sufficient to start the hydrocarbon feed slowly and to allow the catalyst to reduce itself.

Regeneration with steam and air is begun in the same way as for steam alone. A small flow of air is later put into the steam to give about one per cent of air by volume, and the exit is checked for CO_2. If conditions appear steady the air flow can be increased, the tube appearance being carefully watched for any signs of overheating. A visual check of the tubes is generally more satisfactory than reliance on instruments, which may not detect a local overheating.

Blanketing of reduced catalyst

Many catalysts, being pyrophoric in their reduced state, must be kept blanketed with an inert atmosphere, when not in use, to exclude air. Nitrogen is the most satisfactory blanket gas, as it is usually readily available with an oxygen content of less than 10 ppm. Higher oxygen levels are tolerable over some catalysts, which can be re-reduced without loss of activity. If the nitrogen contains more than 100 ppm of oxygen, care is required, since the catalyst may be damaged by rapid oxidation. In such cases the catalyst should be cooled below 50°C before introducing the oxygen-contaminated nitrogen, and the flow of nitrogen should be kept to a minimum to avoid putting more oxygen than is necessary over the catalyst.

During shutdowns, natural gas may be a convenient blanket for the exclusion of air, but its use involves a fire hazard. It can be employed up to temperatures of 300°C without risk of carbon formation.

When a vessel is being opened up whilst under a nitrogen blanket, careful thought must be given to the blanketing arrangements at each stage. If there is more than one opening, there is a danger that the vessel will act as a chimney, and draw air inwards, even though nitrogen is still being supplied. A good general rule is only to have one opening to the atmosphere at any one time.

Ammonia converters require a detailed study to protect the catalyst during maintenance work. Maintenance work is often done at the top and bottom of the converter at the same time, and it is important to ensure that a positive purge is maintained through the catalyst at all stages of the operation. Open ends must not occur simultaneously at the top and bottom of the cartridge. A set of expanding rubber bungs, or inflatable bladders, can be used to close open pipes. Special sealing discs are provided for many ICI converters. The cartridge can be removed from the pressure vessel after all open ends have been closed, and a long nitrogen flex has been firmly attached. It is essential to see that the flex is long enough to remain fixed throughout the lifting operation, and that it will not foul any of the surrounding plant.

When the vessels contain nitrogen, or some other blanketing gas, it is most important to ensure that no person enters without proper breathing equipment. Warning notices should be put at all open manholes.

Several accidents have occured, some of them fatal, through ignorance of the danger courted by entering a vessel containing insufficient oxygen. As the danger is invisible it is easily disregarded. Breathing equipment may be provided, but it will be useless unless everybody involved is aware of the danger, and of the necessity for using the apparatus.

Catalyst discharge

The discharge of catalysts from simple vessels provides few problems if the catalyst is not to be re-used. The catalyst must first be cooled below 50°C, preferably to ambient temperatures. If there is a bottom discharged manhole, a chute will be required to lead the catalyst away from the vessel into drums, or into a tipper truck, or merely to distribute it on the ground. Some catalysts, such as high-temperature shift, low-temperature shift, cobalt/molybdenum and ammonia synthesis catalysts, are pyrophoric in the reduced state, so care is required when they are discharged. Within the vessel the catalyst must be kept blanketed with an inert gas, and upon discharge it must be collected in a metal container on the ground, and away from inflammable materials. If the discharged catalyst becomes hot it can be sprayed with water. Another technique is to discharge the catalyst into drums or a metal trailer containing a few lumps of solid carbon dioxide. This provides a blanketing gas as it evaporates. In some cases the catalyst can be soaked with water before discharge. The water assists in cooling, and also slows down the oxidation of the catalyst material. But this technique can be messy and, when fines are present in the bed, can make discharge more difficult.

It may be preferable to stabilise the catalyst before it is extracted from the vessel. Catalyst which is to be re-used must be stabilised carefully to avoid loss of activity. When the catalyst is to be discarded, more rapid stabilisation is possible, but care is still required.

The following methods have been used for stabilising ICI catalysts.

High-temperature shift catalyst stabilisation

METHOD I—If the catalyst is to be discarded, and has been operating in a sulphur free atmosphere.
  (1) Cool the catalyst to 350–375°C with steam at a space velocity of about 1000 h^{-1}.
  (2) Add between three and five per cent by volume of air to the steam. To avoid possible local overheating and damage to the converter, the measured catalyst temperature must be kept below 500°C. Increase the air rate slightly if this can be done without exceeding the catalyst temperature limits.
  (3) When the catalyst has been oxidised, the steam may be shut off, and the catalysts cooled down by a current of air.

METHOD II—If the catalyst has been operating in the sulphided state, that is, as FeS, the sulphur must be removed from the catalyst at 400–500°C and at a space velocity of up to 1000 h^{-1}. The steam temperature should not be less than 350°C, and steaming should be continued until the concentration of H_2S

in the steam has been reduced to less than 80–100 ppm by volume. This process may take between two and seven days. The catalyst should then be cooled in steam to 200°C and stabilised with air and steam, either as in method I (if the catalyst is to be discarded), or as in method III if it is to be re-used.

METHOD III—If the catalyst is to be re-used the stabilisation is similar to method I, but a more stringent control of temperature is required to avoid loss of activity. The procedure is
  (1) Cool the catalyst to 200°C with steam. Add one per cent by volume of air to the steam to start the oxidation and observe the temperature rise across the converter which should be approximately 30°C.
  (2) After about half an hour, if the temperature rise appears to be normal, increase the air rate to give three per cent of air in steam. After three hours, if the temperature rise is still normal (80–90°C), increase the air rate to a maximum of five per cent by volume of air in steam. During the oxidation the catalyst temperature must not exceed 300°C.
  (3) Analyse the dry exit gas for oxygen to confirm that the oxygen is being consumed. When the oxygen content of the dry exit gas has reached 10 per cent, the concentration of air in the steam can be increased to 10 per cent.
  (4) When the oxidation is complete, there will be no temperature rise in the catalyst, and the oxygen content of the dry exit gas will approach 20 per cent. The steam flow can then be stopped, and the catalyst can be cooled down with air.

Low-temperature shift catalyst stabilisation
  (1) Cool the catalyst to 200°C in nitrogen.
  (2) Add about 0·1 per cent of oxygen to the nitrogen, carefully controlling the temperature so that it does not exceed 220°C.
  (3) When the temperatures have steadied out at about 200°C, slowly increase the oxygen concentration until the catalyst is stable at 200°C with 3 per cent oxygen in the inlet gas.
  (4) Allow the catalyst bed to cool to room temperature with the same gas mixture and slowly increase the oxygen concentration until only air is present in the catalyst bed.

If the catalyst is to be discarded the same procedure can be followed, although the temperature can be allowed to rise above 220°C, so long as the vessel is not overheated.

Methanation catalyst stabilisation
  (1) Cool to 70°C, and purge the converter with inert gas (nitrogen, for example).
  (2) Maintain the flow of inert gas with a space velocity of about 1000 h^{-1}, and add sufficient air to give an oxygen concentration of about five per cent at the inlet.
  (3) Maintain the maximum bed temperature below 250°C by adjusting the oxygen concentration, which must be decreased if there is any indication of temperatures in excess of 250°C.

(4) A hot spot will travel down the bed as the air is added. When temperatures settle down again, more air should be added in order to continue the oxidation, and the process is repeated until no further temperature rise takes place upon the introduction of additional air, or until the oxygen content of the inlet and exit gases is the same.

SAFETY NOTE

Gases containing carbon monoxide must be excluded from the methanator at any temperature below 150°C. Nickel carbonyl is formed below 150°C, and this is an extremely toxic gas.

Ammonia synthesis catalyst stabilisation
(1) Cool the catalyst below 60°C and purge free from hydrogen and ammonia, using nitrogen which should contain less than 100 ppm of oxygen.
(2) Start to circulate the nitrogen at between 3–5 kg cm^{-2} pressure, using the maximum possible circulation rate.
(3) Add air carefully to the circulating mixture to give about 500 ppm of oxygen at the converter inlet, and gradually increase this to 0·3 per cent. Ensure that the maximum temperature remains below 100°C.
(4) When the catalyst is stable with 0·3 per cent oxygen in the inlet gas, steadily increase the oxygen concentration to about seven per cent, continuing to keep the temperature below 95°C.

Discharge of secondary reformers

Secondary reformers are normally discharged from the top, because of the refractory lining of the vessel and the absence of manholes. The top of the catalyst is often protected with a layer of fused alumina chips or bricks, which must be lifted out manually before reaching the catalyst. The catalyst can then be extracted with a vacuum ejector, or can be dug out by hand, using buckets to pull the catalyst out of the vessel. Two men shovelling into large buckets can usually remove the catalyst faster than a vacuum ejector. Dust masks are required, and a compressed air hose is useful to assist in clearing the atmosphere in the vessel. Boards should be provided in the vessel to reduce the amount of trampling on the catalyst.

Discharge of tubular reactors

In catalyst tubes with a flange at the bottom, the catalyst can be discharged easily after the flange has been removed and the catalyst support taken out. Catalyst breakage is reduced if a sock is fitted over the bottom of the tube to lead the catalyst into the drum. It is sometimes necessary to hammer the bottom flange with a hide-faced hammer to loosen any bridging in the tube. If the catalyst is to be re-used it is useful to collect it in three parts corresponding to the bottom, middle and top portions of the tube. The condition of the catalyst can vary with its position in the tube, and it is often possible to recharge part of the catalyst, even if some has to be rejected. Catalyst that has suffered severe breakdown, following a loss of reaction steam, sometimes has to be loosened with augers before it can be discharged.

Tubular reactors with only a top flange are normally discharged with a vacuum extractor. There are several possible arrangements for such equipment. A large industrial vacuum cleaner such as that made by Bivac is often used. As these cleaners are large and awkward to use in a confined space, the vacuum gear is usually left on the ground and connected to a forked manifold with one branch above each row of tubes. A short flexible pipe then connects the manifold to the extraction tube. The extracted catalyst is separated in a cyclone fitted into a hopper or drum, and is then discharged through a chute into a trailer, or into further drums for retention. Compressed air ejectors can be used to provide the vacuum, if a suitable air supply is available. When several tubes are being discharged at once it is better to have an individual vacuum system for each, since a common system does not provide such a good suction.

Even on reformers fitted with bottom flanges, vacuum extraction gear can be useful for discharging part of the catalyst. For instance, when the top of the catalyst is contaminated with boiler solids, it can be removed to leave the remainder in good condition. The catalyst can be discharged a few feet at a time, and the pressure drop of the remainder checked with the device shown in fig. 80. When sufficient catalyst has been removed to give a normal pressure drop, the tube can then be filled up with new catalyst, the remainder being left undisturbed.

Discharge of ammonia synthesis catalyst

Ammonia synthesis catalyst may be stabilised before discharge as described on p. 178. The stabilisation tends to be a long process, requiring special instruments and equipment, but simpler methods have been employed successfully. A cold-shot converter may be discharged directly on to the ground if it has a suitable discharge port. The undischarged catalyst remaining in the vessel must, of course, be blanketed with nitrogen. The ICI 'lozenge' cold-shot converter is particularly suited to being directly discharged, as it has a manhole which allows the catalyst to be discharged without any disturbance to the cartridge itself.

The quantity of catalyst to be disposed of can be a problem in the case of large quench converters. The catalyst cannot simply be poured on to the ground near the converter and some arrangements are needed to remove it during discharging. The unstabilised catalyst can be handled safely so long as it is kept wet. If a water spray is fitted into the discharge chute, the catalyst can be run directly on to a conveyor and from there into a metal lorry. The catalyst is not stabilised by the water and if it is allowed to dry out it will again become pyrophoric. Some hydrogen is evolved by a reaction between the water and the reduced catalyst. If steam from a wet part of the catalyst passes through another part which has become dry and hot, the hydrogen can then ignite. In practice, the discharge can be done safely and easily by providing sprays to keep the catalyst wet but the discharged material should be spread onto open ground in a thin layer as soon as possible.

Most tube-cooled converters, and other types, have to be discharged with vacuum extraction gear through the top cover. When ample supplies of nitrogen are available there is no difficulty, since the catalyst can always be blanketed by introducing nitrogen into the bottom of the converter whilst the extraction is in progress. If, however, the nitrogen supply does not match the rate of extraction, some air will be drawn down through the top of the basket and into the top of the catalyst. This will cause local heating, which may affect the discharge apparatus. In any case, the operators should be provided with insulated gloves for handling the extraction tube. A watch must be kept on the temperature of the catalyst in the bed, and if there are signs of overheating the extraction should be stopped while the nitrogen blanket is re-established. It is also important that the nitrogen/air mixture is not sucked through the discharged catalyst in the extraction hopper. But this can easily be stopped by a purge of nitrogen directed through the bottom of the discharged catalyst. A circulating extraction system is sometimes used when there is a shortage of nitrogen. This requires a circulating pump which sucks the catalyst out from the converter, separates the catalyst from the circulating gas, and returns the gas into the top of the cartridge from the delivery side of the pump. A purge of pure nitrogen is maintained on the undischarged catalyst, and this is supplemented at the top of the bed by the gas returned from the pump delivery.

Re-use of discharged catalyst

If a catalyst is performing satisfactorily it is generally best to leave it undisturbed, and blanketed with an inert gas during shut-down periods. Most catalysts become weakened in use, and discharging and recharging inevitably means loss of material. Reforming catalysts are discharged and recharged both easily and without loss of activity, but some of the material will be lost in the course of extraction and handling. Zinc oxide sulphur removal catalyst can also be discharged and re-used if necessary, without loss of activity. Some check on strength and composition is essential before recharging, and ICI will do this for its customers. Sieving (or even hand-sorting) may be necessary before the recharge, and suitable apparatus must be available. Hand-sorting is made easier by the use of a short conveyor band.

High-temperature shift catalyst and methanation catalyst may be stabilised by controlled oxidation as described in the previous section. These catalysts can be recommissioned without loss of activity by the normal reduction procedure.

Low-temperature shift catalyst may be stabilised for inspection or re-use as described above, but very careful temperature control is required to avoid loss of activity, both during stabilisation, and during the subsequent re-reduction. Stabilisation with a view to re-use is not recommended, since the outcome is rarely satisfactory. Ammonia synthesis catalyst may be stabilised for re-use, but some loss of activity is inevitable, and, again, the practice is not recommended.

Disposal of used catalyst

Catalysts containing nickel, copper, zinc and precious metals may be sold after use to firms dealing in the recovery of such materials. Catalyst which has been discharged in the reduced state must be allowed to become stable in air before dispatch. When the catalyst can be discharged directly into a trailer or old drums, handling costs are reduced. High-temperature shift and ammonia synthesis catalysts are generally scrapped after use, and if discharged in a reduced state, they should be spread out thinly on the ground, and away from any inflammable material, in order to avoid the danger of fires during the oxidation period. If the material is tipped in a heap, the interior can become very hot, so it is important that discharged catalyst should not be tipped together with other waste.

Safety precautions

There are few hazards associated with the use of catalysts, but the following points should be borne in mind.
(1) Catalysts containing metallic nickel must not be exposed to gases containing carbon monoxide at temperatures below 150°C. Observation of this rule avoids the risk of the formation of nickel carbonyl, an extremely toxic, odourless gas which is stable at low temperatures. It is most likely to be formed in methanation reactors when the plant is cooled down, unless the system is thoroughly purged with nitrogen.
(2) Physical hazards arise from the handling of drums, materials, and lifting equipment. These have been outlined in the section on drum handling.
(3) Anybody, not wearing a breathing apparatus, who entered a catalyst vessel filled with an inert gas, would lose consciousness within seconds, and die within minutes. The openings into such vessels must be kept closed, except when work is going on inside them, and, while the vessels are in this condition, prominent warning notices must be displayed. Everybody working within the area should be made aware of the nature and the dangers of asphyxia, and should know how to attempt the rescue and resuscitation of anybody overcome.
(4) Catalysts discharged in the pyrophoric state must be kept away from inflammable materials. Dumps of such pyrophoric waste should be within reach of water hoses, since overheating may occur. High temperatures can build up in the hearts of heaps of discharged catalysts, so that the material is best spread thinly over the ground, and until the heating reactions are finished, it should not be walked upon.
(5) Some catalysts (high-temperature shift agents, and some used in methanation) contain chrome. Dust masks should be provided for people working with such materials.

Indeed, dust masks can be worn with advantage by anybody working in a dusty atmosphere, and particularly by people working inside closed catalyst vessels.

But it must be made clear that a dust mask is not a breathing apparatus, and is useless as a protection against poisonous gases, or in an atmosphere lacking oxygen.

General references on catalysis

ELEY, D. D., PINES, H., *and* WEISS, P. B. (Eds) *Advances in Catalysis and Related Subjects*, Vols. 1–18 et seq. Academic Press, New York and London.

EMMETT, PAUL H. (Ed.) *Catalysis*, Vols. 1–7. Reinhold Publishing Corpn, New York (1960).

THOMAS, J. M. *and* THOMAS, W. J. *Introduction to the Principles of Heterogeneous Catalysis*. Academic Press, New York and London (1967).

BOND, G. C. *Catalysis by Metals*. Academic Press, New York and London (1962).

References

1. SCHÄFER, H, *Chemical Transport Reactions*. Academic Press, New York (1964).
2. HAYWOOD, D. O. *and* TRAPNELL, B. M. W., *Chemisorption*. Butterworth, London (1964).
3. SCHÄCHTER, Y. *and* PINES, H. *J. Catalysis*, **11**, 147 (1968).
4. DOWDEN, D. A. 4th International Congress on Catalysis, Moscow, 1968. Plenary Lecture.
5. SELWOOD, P. W. *J. Amer. Chem. Soc.* **87**, 1804 (1965); **88**, 2676 (1966).
6. WEISZ, P. W. *Advances in Catalysis*, Vol. 13, 137 (1962).
7. EMMETT, P. H. *Advances in Catalysis*, Vol. 1 (1948).
8. WHEELER, A. *Catalysis*, Vol. 2 (1955).
9. THIELE, E. W. *Ind. Eng. Chem.* **31**, 316 (1939).
10. ROBLEE, L. H. S., BAIRD, R. M., *and* TIERNEY, J. W. *A.I.Ch.E. Journal* **4**, 461 (Dec. 1958).
11. SCHWARTZ, C. E., *and* SMITH, J. M. *Ind. Eng. Chem.* **45**, 1209 (1953).
12. BORESKOV, G. K. *Kinetics and Catalysis USSR* (English translation), **3**, 416 (July–Aug. 1962).
13. CARBERRY, J. J., *and* GILLESPIE, B. M. *Ind. Eng. Chem.* **58**, 164 (1966).
14. BRISK, M. L., DAY, R. L., JONES, M., *and* WARREN, J. B. *Trans. Int. Chem. Eng.* **46**, T3 (1968).
15. *Identification of Sulphur Compounds in Petroleum*. US Bureau of Mines, R.I. 6803 (1966).
16. RICHARDSON, J. T. *Ind. and Eng. Chem. Fundamentals*, **3**, 154 (1964).
17. REID, E. E. *Organic Chemistry of Bivalent Sulphur*, Vol. 2. Chemical Publishing Co. Inc. (1959).
18. RUDENKO, M. G., *and* GROMOVA, U. N. *Chemical Abstracts* **46**, 7515d (1952).
19. MCKINLEY, J. B. *Catalysis* (ed. P. H. Emmett), Vol. 5. Reinhold (1957).
20. DESIKAN, P., *and* AMBERG, G. H. *Canadian J. Chem.* **42**, 843 (1964).
21. PHILLIPSON, J. J. Unpublished work.
22. GHOSHAL, S. R., *et al. Technology* **3** (3), 126 (1966).
23. GHOSHAL, S. R., *et al. Technology* **3** (1), 3 (1966).
24. GHOSHAL, S. R., *et al. Technology* **2** (4), 211 (1965).
25. WILSON, W. A., VOREK, W. E., *and* MALO, R. V. *Ind. Eng. Chem.* **49** (4), 657 (1957).
26. TADAO OHTSUKA, *et al. Bull. Japan Pet. Institute* **2**, 13 (1960).
27. OWENS, P. J., *and* AMBERG, C. H. *Advances in Chemistry Series* **33**, 182 (1961).
28. BRIDGER, G. W., *and* WYRWAS, W. *Chem. Proc. Eng.* **48**, 101 (1967).
29. DENT, F. J. 49th Report of the Joint Research Committee, *Trans. Inst. Gas Eng.* (1945–6).
30. GILLILAND, E. R., *and* HARRIOTT, P. *Ind. Eng. Chem.* **46**, 2195 (1954).
31. MURPHY, D. B., *et al. Proc. Conf. on Industrial Carbon and Graphite* (Soc. Chem. Ind.), 77 (1957).
32. GULBRANSEN, E. A., *and* ANDREW, K. F. *Ind. Eng. Chem.* **44**, 1034 (1952).
33. GORING, G. E., *et al. Ind. Eng. Chem.* **44**, 1051 (1952).
34. HEDDEN, K. *Proc. 5th Carbon Conf.*, 125 (1961).
35. GADSBY, J., *et al. Proc. Roy. Soc.* **A187**, 129 (1946).
36. DENT, F. J. *Gas J.* **184**, 199 (1928); STANIER, H., *and* MCKEAN, *Gas J.* **265**, 111 (1951).
37. HARKER, H. *Proc. 4th Carbon Conf.*, p. 95 (1959); EARP, F. W., *and* HILL, M. W. *Proc. Conf. on Industrial Carbon and Graphite* (Soc. Chem. Ind.), 326 (1957).
38. ICI, British Patent 1,029,235.
39. ICI, French Patent 1,559,218.

REFERENCES

40. ICI, British Patent 1,032,752.
41. ICI, German Patent P. 1,792,523.
42. ICI, South African Patent 5863/68.
43. RIESZ, C. H., LURIE, P. C., TSAROS, C. L., and PETTYJOHN, E. S. *Institute of Gas Technology Research Bulletin* No. 6.
44. RIESZ, C. H., DIRKSEN, H. A., and KIRKPATRICK, W. J. *Institute of Gas Technology Research Bulletin* No. 10.
45. RIESZ, C. H., DIRKSEN, H. A., and PLETICKA, W. J. *Institute of Gas Technology Research Bulletin* No. 20.
46. ARNOLD, M. R., ATWOOD, K., BAUGH, H. M., and SMYSER, H. D. *Ind. Eng. Chem.* **44**, No. 5, 999 (1952).
47. FEITSMA, *Mitt. V.G.B.* **72**, 170 (June 1961).
48. STRAUB, F. G. *University of Illinois Bulletin* **43**, No. 59 (1946).
49. ICI, British Patent 953,877.
50. ICI, British Patent 1,003,702.
51. ICI, British Patent 1,040,066.
52. KELLOGG, British Patent 966,882.
53. KELLOGG, British Patent 966,883.
54. MORITA, S., and INOUE, T. *Int. Chem. Eng.* **5**, No. 1, 180 (1965).
55. PICHLER, H. *Advances in Catalysis*, Vol. 4, 326.
56. KIRKPATRICK, W. J. *Advances in Catalysis*, Vol. 3, 331.
57. BODROV, N. M., APEL'BAUM, L. O., and TEMKIN, M. I. *Kinetics and Catalysis* (English translation) **5**, No. 4, 614–22 (1964).
58. AKERS, W. W., and CAMP, D. P. *A.I. Chem. Eng. Journal* **4**, 471 (1955).
59. WRIGHT, P. G., ASHMORE, P. G., and KEMBALL, C. *Trans. Faraday Soc.* **54**, 1692 (1958).
60. KEMBALL, C. *Proc. Roy. Soc.* **A207**, 529 (1951); **A217**, 376 (1953).
61. BODROV, N. M., APEL'BAUM, L. O., and TEMKIN, M. I. *Kinetics and Catalysis* (English translation) **8**, No. 4, 696–702 (1967).
62. UCHIDA, H., ISOGAI, N., OBA, M., and HASEGAWA, T. *Bull. Chem. Soc. Japan* **40** (8), 1981–6 (1967).
63. MOE, J. M. Symposium on Production of Hydrogen, Division of Petroleum Chemistry, American Chemical Soc. Sept. 1963, B.29.
64. SHCHIBRYA, G. G., MOROZOV, N. M., and TEMKIN, M. I. *Kinetics and Catalysis* **6**, No. 6 (1965).
65. CAMPBELL, J. S., and METCALFE, S. R. To be published.
66. HABER, F., TAMARU, S., and PONNAZ, C. *Z. Elektrochem.* **21**, 89 (1915).
67. LARSON, A. T., and DODGE, R. L. *J. Am. Chem. Soc.* **45**, 2918 (1923).
68. GILLESPIE, L. J., and BEATTIE, J. A. *Phys. Rev.* **36**, 743 (1930).
69. NIELSEN, A. *An Investigation on Promoted Iron Catalysts for the Synthesis of Ammonia*, 20. J. Gjellerups Forlag, Copenhagen (1968).
70. TEMKIN, M. I., and PYZHEV, V. *Acta. Physicochim.* (*U.R.S.S.*) **12**, 327 (1940).
71. ANNABLE, D. *Chem. Eng. Sci.* **1**, 145 (1952).
72. LIVSHITS, V. D., and SIDOROV, I. P. *Zhur. Fiz. Khim.* **26**, 538 (1952).
73. TEMKIN, M. I., MOROZOV, N. M., and SHAPATINA, E. N. *Kinetics and Catalysis* (*U.R.S.S.*) **4**, 565 (1963).
74. TEMKIN, M. I., MOROZOV, N. M., and SHAPATINA, E. N. *Kinetics and Catalysis* (*U.R.S.S.*) **4**, 260 (1963).
75. TEMKIN, M. I., KIPERMAN, S. L., and LUK'YANOV, L. I. *Dokl. AN.S.S.S.R.* **74**, 763 (1950).
76. SPENDLEY, W., HEXT, G. R., and HIMSWORTH, F. R. *Technometrics* **4**, No. 4, 441–461, Nov. 1962.
77. HORN, F. *Zeitschrift Für Elektrochemie* **65**, 295 (1961).

APPENDICES

TABLES 1–16

TABLE 1
GENERAL PROPERTIES OF ICI CATALYSTS FOR AMMONIA PROCESSES

ICI Catalyst No.	Duty	Form	Size	Composition	Bulk density
46-1	Primary reforming	Raschig rings	17 × 17 mm	Supported nickel oxide	1·15
57-1	Primary reforming	Raschig rings	17 × 17 mm	Supported nickel oxide	1·1
54-2	Secondary reforming	Cylindrical pellets	17 × 17 mm	Supported nickel oxide	1·0
15-4	High-temperature shift	Cylindrical pellets	8·5 × 11·3 mm	Iron oxide chromia	1·35
15-5	High-temperature shift	Cylindrical pellets	5·4 × 3·6 mm	Iron oxide chromia	1·35
52-1	Low-temperature shift	Cylindrical pellets	5·4 × 3·6 mm	Oxides of copper, zinc, alumina	0·8–0·9
11-3	Methanation	Cylindrical pellets	5·4 × 3·6 mm	Supported nickel oxide	1·0–1·1
35-4	Ammonia synthesis	Granules	3–9 mm	Promoted magnetite	2·6–2·9
32-4	Desulphurisation	Spheres	$\frac{1}{8}-\frac{3}{16}$ in	Zinc oxide	1·1
41-4	Desulphurisation	Extrusions	$\frac{1}{8}$ in dia.	Oxides of cobalt, molybdenum, alumina	0·8–0·9
41-3	Desulphurisation	Pellets	5·4 × 3·6 mm	Oxides of cobalt, molybdenum, alumina	1·1

TABLE 2
PERIODIC TABLE OF THE ELEMENTS

	Group I	Group II	Group III	Group IV	Group V	Group VI	Group VII	Group VIII			Gp. O
0											0 Nn 1·0088
1	1 H 1·00797										2 He 4·0026
2	3 Li 6·939	4 Be 9·0122	5 B 10·811	6 C 12·01115	7 N 14·0067	8 O 15·9994	9 F 18·9984				10 Ne 20·183
3	11 Na 22·9898	12 Mg 24·312	13 Al 26·9815	14 Si 28·086	15 P 30·9738	16 S 32·064	17 Cl 35·453				18 A 39·948
4	19 K 39·102	20 Ca 40·08	21 Sc 44·956	22 Ti 47·90	23 V 50·942	24 Cr 51·996	25 Mn 54·9381	26 Fe 55·847	27 Co 58·933	28 Ni 58·71	
	29 Cu 63·54	30 Zn 65·37	31 Ga 69·72	32 Ge 72·59	33 As 74·9216	34 Se 78·96	35 Br 79·909				36 Kr 83·80
5	37 Rb 85·47	38 Sr 87·62	39 Y 88·905	40 Zr 91·22	41 Nb 92·906	42 Mo 95·94	43 Tc	44 Ru 101·07	45 Rh 102·905	46 Pd 106·4	
	47 Ag 107·870	48 Cd 112·40	49 In 114·82	50 Sn 118·69	51 Sb 121·75	52 Te 127·60	53 I 126·9044				54 Xe 131·30
6	55 Cs 132·905	56 Ba 137·34	57 La 138·91	72 Hf 178·49	73 Ta 180·948	74 W 183·85	75 Re 186·2	76 Os 190·2	77 Ir 192·2	78 Pt 195·09	
	79 Au 196·967	80 Hg 200·59	81 Tl 204·37	82 Pb 207·19	83 Bi 208·980	84 Po	85 At				86 Rn
7	87 Fr	88 Ra	89 Ac	90 Th 232·038	91 Pa	92 U 238·03					

Lanthanides	57 La	58 Ce	59 Pr	60 Nd	61 Pm	62 Sm	63 Eu	64 Gd	65 Tb	66 Dy	67 Ho	68 Er	69 Tm	70 Yb	71 Lu
												167·26	168·934	173·04	174·97

TABLE 3

DESULPHURISATION OVER ZINC OXIDE

Gas phase equilibrium constants

$$ZnO + H_2S \rightleftharpoons ZnS + H_2O$$

Table 3a **Table 3b**

Temperature (°C)	$K_p = \dfrac{P_{H_2O}}{P_{H_2S}}$	Temperature (°F)	$K_p = \dfrac{P_{H_2O}}{P_{H_2S}}$
200	$2\cdot081 \times 10^8$	400	$1\cdot738 \times 10^8$
220	$9\cdot494 \times 10^7$	425	$1\cdot011 \times 10^8$
240	$4\cdot605 \times 10^7$	450	$6\cdot061 \times 10^7$
260	$2\cdot359 \times 10^7$	475	$3\cdot734 \times 10^7$
280	$1\cdot268 \times 10^7$	500	$2\cdot359 \times 10^7$
300	$7\cdot121 \times 10^6$	525	$1\cdot526 \times 10^7$
320	$4\cdot157 \times 10^6$	550	$1\cdot008 \times 10^7$
340	$2\cdot514 \times 10^6$	575	$6\cdot800 \times 10^6$
360	$1\cdot569 \times 10^6$	600	$4\cdot671 \times 10^6$
380	$1\cdot008 \times 10^6$	625	$3\cdot265 \times 10^6$
400	$6\cdot648 \times 10^5$	650	$2\cdot319 \times 10^6$
420	$4\cdot491 \times 10^5$	675	$1\cdot672 \times 10^6$
440	$3\cdot101 \times 10^5$	700	$1\cdot223 \times 10^6$
460	$2\cdot185 \times 10^5$	725	$9\cdot063 \times 10^5$
480	$1\cdot568 \times 10^5$	750	$6\cdot799 \times 10^5$
500	$1\cdot145 \times 10^5$	775	$5\cdot161 \times 10^5$
		800	$3\cdot960 \times 10^5$
		825	$3\cdot070 \times 10^5$
		850	$2\cdot404 \times 10^5$
		875	$1\cdot899 \times 10^5$
		900	$1\cdot513 \times 10^5$
		925	$1\cdot216 \times 10^5$
		950	$9\cdot844 \times 10^4$

TABLE 4

METHANE–STEAM REACTION

Gas phase equilibrium constants

$$CH_4 + H_2O \rightleftharpoons CO + 3H_2$$

Table 4a

Temperature (°C)	$K_p = \dfrac{P_{CO} P_{H_2}^3}{P_{CH_4} P_{H_2O}}$	Increment/°C	Temperature (°C)	$K_p = \dfrac{P_{CO} P_{H_2}^3}{P_{CH_4} P_{H_2O}}$	Increment/°C
200	4.614×10^{-12}		760	6.149×10^1	0.166×10^1
250	8.397×10^{-10}		765	6.981×10^1	0.187×10^1
300	6.378×10^{-8}		770	7.917×10^1	0.210×10^1
350	2.483×10^{-6}		775	8.968×10^1	0.024×10^2
400	5.732×10^{-5}		780	1.015×10^2	0.026×10^2
450	8.714×10^{-4}		785	1.147×10^2	0.030×10^2
500	9.442×10^{-3}		790	1.294×10^2	0.033×10^2
550	7.741×10^{-2}		795	1.459×10^2	0.037×10^2
600	5.029×10^{-1}		800	1.644×10^2	0.041×10^2
605	5.996×10^{-1}	0.193×10^{-1}	805	1.849×10^2	0.046×10^2
610	7.135×10^{-1}	0.228×10^{-1}	810	2.079×10^2	0.051×10^2
615	8.474×10^{-1}	0.268×10^{-1}	815	2.334×10^2	0.057×10^2
620	1.005	0.031	820	2.617×10^2	0.063×10^2
625	1.189	0.037	825	2.933×10^2	0.070×10^2
630	1.404	0.043	830	3.282×10^2	0.078×10^2
635	1.656	0.050	835	3.670×10^2	0.086×10^2
640	1.949	0.059	840	4.100×10^2	0.095×10^2
645	2.290	0.068	845	4.576×10^2	0.105×10^2
650	2.686	0.079	850	5.101×10^2	0.116×10^2
655	3.145	0.092	855	5.682×10^2	0.128×10^2
660	3.677	0.106	860	6.323×10^2	0.141×10^2
665	4.293	0.123	865	7.030×10^2	0.156×10^2
670	5.003	0.142	870	7.809×10^2	0.171×10^2
675	5.821	0.164	875	8.666×10^2	0.189×10^2
680	6.763	0.188	880	9.609×10^2	0.021×10^3
685	7.845	0.216	885	1.064×10^3	0.023×10^3
690	9.087	0.248	890	1.178×10^3	0.025×10^3
695	1.051×10^1	0.028×10^1	895	1.303×10^3	0.027×10^3
700	1.214×10^1	0.033×10^1	900	1.440×10^3	0.030×10^3
705	1.400×10^1	0.037×10^1	905	1.589×10^3	0.033×10^3
710	1.612×10^1	0.042×10^1	910	1.753×10^3	0.036×10^3
715	1.854×10^1	0.048×10^1	915	1.933×10^3	0.039×10^3
720	2.129×10^1	0.055×10^1	920	2.128×10^3	0.043×10^3
725	2.442×10^1	0.063×10^1	925	2.342×10^3	0.047×10^3
730	2.797×10^1	0.071×10^1	930	2.575×10^3	0.051×10^3
735	3.200×10^1	0.081×10^1	935	2.830×10^3	0.055×10^3
740	3.656×10^1	0.091×10^1	940	3.107×10^3	0.060×10^3
745	4.171×10^1	0.103×10^1	945	3.408×10^3	0.066×10^3
750	4.753×10^1	0.116×10^1	950	3.736×10^3	0.071×10^3
755	5.409×10^1	0.131×10^1	955	4.092×10^3	0.077×10^3
		0.148×10^1			

Temperature (°C)	$K_p = \dfrac{P_{CO} P_{H_2}^3}{P_{CH_4} P_{H_2O}}$	Increment/°C
960	4.480×10^3	0.084×10^3
965	4.900×10^3	0.091×10^3
970	5.356×10^3	0.099×10^3
975	5.850×10^3	0.107×10^3

Temperature (°C)	$K_p = \dfrac{P_{CO} P_{H_2}^3}{P_{CH_4} P_{H_2O}}$	Increment/°C
980	6.385×10^3	0.116×10^3
985	6.964×10^3	0.125×10^3
990	7.590×10^3	0.135×10^3
995	8.268×10^3	0.146×10^3

Table 4b

Temperature (°F)	$K_p = \dfrac{P_{CO} P_{H_2}^3}{P_{CH_4} P_{H_2O}}$	Increment/°F
400	7.640×10^{-12}	
500	2.124×10^{-9}	
600	2.123×10^{-7}	
700	9.891×10^{-6}	
800	2.563×10^{-4}	
900	4.189×10^{-3}	
1000	4.736×10^{-2}	
1100	3.966×10^{-1}	0.087×10^{-1}
1110	4.835×10^{-1}	0.105×10^{-1}
1120	5.880×10^{-1}	0.125×10^{-1}
1130	7.134×10^{-1}	0.150×10^{-1}
1140	8.635×10^{-1}	0.018
1150	1.043	0.021
1160	1.257	0.025
1170	1.511	0.030
1180	1.813	0.036
1190	2.170	0.042
1200	2.592	0.050
1210	3.091	0.059
1220	3.677	0.069
1230	4.366	0.081
1240	5.174	0.095
1250	6.120	0.110
1260	7.225	0.129
1270	8.513	0.015×10^1
1280	1.001×10^1	0.017×10^1
1290	1.176×10^1	0.020×10^1
1300	1.378×10^1	0.023×10^1
1310	1.612×10^1	0.027×10^1
1320	1.883×10^1	0.031×10^1
1330	2.195×10^1	0.036×10^1
1340	2.555×10^1	0.041×10^1
1350	2.970×10^1	0.048×10^1
1360	3.446×10^1	0.055×10^1
1370	3.992×10^1	0.063×10^1
1380	4.617×10^1	0.072×10^1
1390	5.332×10^1	0.082×10^1
1400	6.148×10^1	0.093×10^1
1410	7.079×10^1	0.106×10^1

Temperature (°F)	$K_p = \dfrac{P_{CO} P_{H_2}^3}{P_{CH_4} P_{H_2O}}$	Increment/°F
1420	8.139×10^1	0.121×10^1
1430	9.344×10^1	0.014×10^2
1440	1.071×10^2	0.016×10^2
1450	1.226×10^2	0.018×10^2
1460	1.402×10^2	0.020×10^2
1470	1.601×10^2	0.022×10^2
1480	1.825×10^2	0.025×10^2
1490	2.078×10^2	0.029×10^2
1500	2.363×10^2	0.032×10^2
1510	2.684×10^2	0.036×10^2
1520	3.045×10^2	0.040×10^2
1530	3.450×10^2	0.045×10^2
1540	3.903×10^2	0.051×10^2
1550	4.411×10^2	0.057×10^2
1560	4.979×10^2	0.063×10^2
1570	5.614×10^2	0.071×10^2
1580	6.323×10^2	0.079×10^2
1590	7.112×10^2	0.088×10^2
1600	7.992×10^2	0.098×10^2
1610	8.970×10^2	0.011×10^3
1620	1.006×10^3	0.012×10^3
1630	1.126×10^3	0.013×10^3
1640	1.260×10^3	0.015×10^3
1650	1.408×10^3	0.016×10^3
1660	1.572×10^3	0.018×10^3
1670	1.753×10^3	0.020×10^3
1680	1.953×10^3	0.022×10^3
1690	2.174×10^3	0.024×10^3
1700	2.418×10^3	0.027×10^3
1710	2.686×10^3	0.029×10^3
1720	2.980×10^3	0.032×10^3
1730	3.305×10^3	0.036×10^3
1740	3.661×10^3	0.039×10^3
1750	4.051×10^3	0.043×10^3
1760	4.479×10^3	0.047×10^3
1770	4.948×10^3	0.051×10^3
1780	5.462×10^3	0.056×10^3
1790	6.023×10^3	0.061×10^3

TABLE 5

WATER-GAS SHIFT REACTION

Gas phase equilibrium constants

$$CO + H_2O \rightleftharpoons CO_2 + H_2$$

Table 5a

Temperature (°C)	$K_p = \dfrac{P_{CO_2} P_{H_2}}{P_{CO} P_{H_2O}}$	Increment/°C	Temperature (°C)	$K_p = \dfrac{P_{CO_2} P_{H_2}}{P_{CO} P_{H_2O}}$	Increment/°C
200	2.279×10^2		400	1.170×10^1	
205	2.049×10^2	-0.046×10^2	405	1.113×10^1	-0.012×10^1
210	1.846×10^2	-0.041×10^2	410	1.059×10^1	-0.011×10^1
215	1.667×10^2	-0.036×10^2	415	1.008×10^1	-0.010×10^1
220	1.509×10^2	-0.032×10^2	420	9.610	-0.095
225	1.369×10^2	-0.028×10^2	425	9.165	-0.089
230	1.244×10^2	-0.025×10^2	430	8.748	-0.083
235	1.133×10^2	-0.022×10^2	435	8.356	-0.078
240	1.034×10^2	-0.020×10^2	440	7.986	-0.074
245	9.447×10^1	-0.178×10^1	445	7.639	-0.070
250	8.651×10^1	-0.159×10^1	450	7.311	-0.066
255	7.936×10^1	-0.143×10^1	455	7.002	-0.062
260	7.293×10^1	-0.129×10^1	460	6.710	-0.058
265	6.712×10^1	-0.116×10^1	465	6.435	-0.055
270	6.189×10^1	-0.105×10^1	470	6.174	-0.052
275	5.714×10^1	-0.095×10^1	475	5.928	-0.049
280	5.285×10^1	-0.086×10^1	480	5.695	-0.047
285	4.894×10^1	-0.078×10^1	485	5.474	-0.044
290	4.540×10^1	-0.071×10^1	490	5.265	-0.042
295	4.217×10^1	-0.065×10^1	495	5.067	-0.040
300	3.922×10^1	-0.059×10^1	500	4.878	-0.038
305	3.652×10^1	-0.054×10^1	505	4.700	-0.036
310	3.406×10^1	-0.049×10^1	510	4.530	-0.034
315	3.180×10^1	-0.045×10^1	515	4.368	-0.032
320	2.973×10^1	-0.041×10^1	520	4.215	-0.031
325	2.783×10^1	-0.038×10^1	525	4.069	-0.029
330	2.608×10^1	-0.035×10^1	530	3.929	-0.028
335	2.447×10^1	-0.032×10^1	535	3.797	-0.027
340	2.298×10^1	-0.030×10^1	540	3.670	-0.025
345	2.161×10^1	-0.027×10^1	545	3.550	-0.024
350	2.034×10^1	-0.025×10^1	550	3.434	-0.023
355	1.916×10^1	-0.024×10^1	555	3.325	-0.022
360	1.807×10^1	-0.022×10^1	560	3.220	-0.021
365	1.706×10^1	-0.020×10^1	565	3.119	-0.020
370	1.612×10^1	-0.019×10^1	570	3.023	-0.019
375	1.525×10^1	-0.017×10^1	575	2.931	-0.018
380	1.444×10^1	-0.016×10^1	580	2.843	-0.018
385	1.368×10^1	-0.015×10^1	585	2.759	-0.017
390	1.298×10^1	-0.014×10^1	590	2.679	-0.016
395	1.232×10^1	-0.013×10^1	595	2.601	-0.015
		-0.012×10^1			-0.015

Temperature (°C)	$K_p = \dfrac{P_{CO_2} P_{H_2}}{P_{CO} P_{H_2O}}$	Increment/°C	Temperature (°C)	$K_p = \dfrac{P_{CO_2} P_{H_2}}{P_{CO} P_{H_2O}}$	Increment/°C
600	2·527	−0·014	800	1·015	−0·036 × 10⁻¹
605	2·456	−0·014	805	9·968 × 10⁻¹	−0·035 × 10⁻¹
610	2·388	−0·013	810	9·793 × 10⁻¹	−0·034 × 10⁻¹
615	2·322	−0·013	815	9·622 × 10⁻¹	−0·033 × 10⁻¹
620	2·259	−0·012	820	9·457 × 10⁻¹	−0·032 × 10⁻¹
625	2·199	−0·012	825	9·295 × 10⁻¹	−0·031 × 10⁻¹
630	2·141	−0·011	830	9·139 × 10⁻¹	−0·031 × 10⁻¹
635	2·085	−0·011	835	8·986 × 10⁻¹	−0·030 × 10⁻¹
640	2·031	−0·010	840	8·837 × 10⁻¹	−0·029 × 10⁻¹
645	1·979	−0·010	845	8·693 × 10⁻¹	−0·028 × 10⁻¹
650	1·923	−0·010	850	8·552 × 10⁻¹	−0·027 × 10⁻¹
655	1·881	−0·009	855	8·415 × 10⁻¹	−0·027 × 10⁻¹
660	1·835	−0·009	860	8·282 × 10⁻¹	−0·026 × 10⁻¹
665	1·790	−0·009	865	8·152 × 10⁻¹	−0·025 × 10⁻¹
670	1·747	−0·008	870	8·025 × 10⁻¹	−0·025 × 10⁻¹
675	1·706	−0·008	875	7·901 × 10⁻¹	−0·024 × 10⁻¹
680	1·666	−0·008	880	7·781 × 10⁻¹	−0·023 × 10⁻¹
685	1·627	−0·007	885	7·664 × 10⁻¹	−0·023 × 10⁻¹
690	1·590	−0·007	890	7·549 × 10⁻¹	−0·022 × 10⁻¹
695	1·554	−0·007	895	7·437 × 10⁻¹	−0·022 × 10⁻¹
700	1·519	−0·007	900	7·328 × 10⁻¹	−0·021 × 10⁻¹
705	1·485	−0·007	905	7·222 × 10⁻¹	−0·021 × 10⁻¹
710	1·453	−0·006	910	7·118 × 10⁻¹	−0·020 × 10⁻¹
715	1·421	−0·006	915	7·017 × 10⁻¹	−0·020 × 10⁻¹
720	1·391	−0·006	920	6·918 × 10⁻¹	−0·019 × 10⁻¹
725	1·361	−0·006	925	6·822 × 10⁻¹	−0·019 × 10⁻¹
730	1·333	−0·006	930	6·728 × 10⁻¹	−0·018 × 10⁻¹
735	1·305	−0·005	935	6·636 × 10⁻¹	−0·018 × 10⁻¹
740	1·279	−0·005	940	6·546 × 10⁻¹	−0·018 × 10⁻¹
745	1·253	−0·005	945	6·458 × 10⁻¹	−0·017 × 10⁻¹
750	1·228	−0·005	950	6·372 × 10⁻¹	−0·017 × 10⁻¹
755	1·203	−0·005	955	6·288 × 10⁻¹	−0·016 × 10⁻¹
760	1·180	−0·005	960	6·206 × 10⁻¹	−0·016 × 10⁻¹
765	1·157	−0·004	965	6·126 × 10⁻¹	−0·016 × 10⁻¹
770	1·135	−0·004	970	6·047 × 10⁻¹	−0·015 × 10⁻¹
775	1·113	−0·004	975	5·971 × 10⁻¹	−0·015 × 10⁻¹
780	1·092	−0·004	980	5·896 × 10⁻¹	−0·015 × 10⁻¹
785	1·072	−0·004	985	5·822 × 10⁻¹	−0·014 × 10⁻¹
790	1·053	−0·004	990	5·750 × 10⁻¹	−0·014 × 10⁻¹
795	1·033	−0·004	995	5·680 × 10⁻¹	−0·014 × 10⁻¹

Table 5b

Temperature (°F)	$K_p = \dfrac{P_{CO_2} P_{H_2}}{P_{CO} P_{H_2O}}$	Increment/°F	Temperature (°F)	$K_p = \dfrac{P_{CO_2} P_{H_2}}{P_{CO} P_{H_2O}}$	Increment/°F
400	2.073×10^2	-0.023×10^2	890	5.849	-0.025
410	1.846×10^2	-0.020×10^2	900	5.596	-0.024
420	1.649×10^2	-0.017×10^2	910	5.357	-0.023
430	1.477×10^2	-0.015×10^2	920	5.132	-0.021
440	1.326×10^2	-0.013×10^2	930	4.919	-0.020
450	1.193×10^2	-0.012×10^2	940	4.719	-0.019
460	1.076×10^2	-0.103×10^1	950	4.530	-0.018
470	9.733×10^1	-0.091×10^1	960	4.351	-0.017
480	8.821×10^1	-0.081×10^1	970	4.182	-0.016
490	8.012×10^1	-0.072×10^1	980	4.021	-0.015
500	7.293×10^1	-0.064×10^1	990	3.870	-0.014
510	6.652×10^1	-0.057×10^1	1000	3.726	-0.014
520	6.079×10^1	-0.051×10^1	1010	3.589	-0.013
530	5.567×10^1	-0.046×10^1	1020	3.460	-0.012
540	5.107×10^1	-0.041×10^1	1030	3.337	-0.012
550	4.694×10^1	-0.037×10^1	1040	3.220	-0.011
560	4.321×10^1	-0.034×10^1	1050	3.108	-0.011
570	3.985×10^1	-0.030×10^1	1060	3.003	-0.010
580	3.681×10^1	-0.028×10^1	1070	2.902	-0.010
590	3.406×10^1	-0.025×10^1	1080	2.806	-0.009
600	3.157×10^1	-0.023×10^1	1090	2.714	-0.009
610	2.930×10^1	-0.021×10^1	1100	2.627	-0.008
620	2.723×10^1	-0.019×10^1	1110	2.543	-0.008
630	2.535×10^1	-0.017×10^1	1120	2.464	-0.008
640	2.363×10^1	-0.016×10^1	1130	2.388	-0.007
650	2.205×10^1	-0.014×10^1	1140	2.315	-0.007
660	2.061×10^1	-0.013×10^1	1150	2.246	-0.007
670	1.929×10^1	-0.012×10^1	1160	2.179	-0.006
680	1.807×10^1	-0.011×10^1	1170	2.116	-0.006
690	1.696×10^1	-0.010×10^1	1180	2.055	-0.006
700	1.592×10^1	-0.010×10^1	1190	1.996	-0.006
710	1.497×10^1	-0.009×10^1	1200	1.940	-0.005
720	1.410×10^1	-0.008×10^1	1210	1.886	-0.005
730	1.328×10^1	-0.008×10^1	1220	1.835	-0.005
740	1.253×10^1	-0.007×10^1	1230	1.785	-0.005
750	1.184×10^1	-0.006×10^1	1240	1.738	-0.005
760	1.119×10^1	-0.006×10^1	1250	1.692	-0.004
770	1.059×10^1	-0.006×10^1	1260	1.648	-0.004
780	1.003×10^1	-0.052	1270	1.606	-0.004
790	9.509	-0.049	1280	1.566	-0.004
800	9.024	-0.045	1290	1.527	-0.004
810	8.571	-0.042	1300	1.489	-0.004
820	8.148	-0.040	1310	1.453	-0.003
830	7.752	-0.037	1320	1.418	-0.003
840	7.382	-0.035	1330	1.384	-0.003
850	7.036	-0.033	1340	1.352	-0.003
860	6.710	-0.031	1350	1.321	-0.003
870	6.405	-0.029	1360	1.290	-0.003
880	6.119	-0.027	1370	1.261	-0.003

Temperature (°F)	$K_p = \dfrac{P_{CO_2} P_{H_2}}{P_{CO} P_{H_2O}}$	Increment/°F	Temperature (°F)	$K_p = \dfrac{P_{CO_2} P_{H_2}}{P_{CO} P_{H_2O}}$	Increment/°F
1380	1·233	−0·003	1590	$8·137 \times 10^{-1}$	$−0·014 \times 10^{-1}$
1390	1·206	−0·003	1600	$7·997 \times 10^{-1}$	$−0·014 \times 10^{-1}$
1400	1·180	−0·003	1610	$7·861 \times 10^{-1}$	$−0·013 \times 10^{-1}$
1410	1·155	−0·002	1620	$7·728 \times 10^{-1}$	$−0·013 \times 10^{-1}$
1420	1·130	−0·002	1630	$7·600 \times 10^{-1}$	$−0·013 \times 10^{-1}$
1430	1·106	−0·002	1640	$7·474 \times 10^{-1}$	$−0·012 \times 10^{-1}$
1440	1·083	−0·002	1650	$7·353 \times 10^{-1}$	$−0·012 \times 10^{-1}$
1450	1·061	−0·002	1660	$7·234 \times 10^{-1}$	$−0·012 \times 10^{-1}$
1460	1·040	−0·002	1670	$7·119 \times 10^{-1}$	$−0·011 \times 10^{-1}$
1470	1·019	$−0·020 \times 10^{-1}$	1680	$7·006 \times 10^{-1}$	$−0·011 \times 10^{-1}$
1480	$9·988 \times 10^{-1}$	$−0·020 \times 10^{-1}$	1690	$6·897 \times 10^{-1}$	$−0·011 \times 10^{-1}$
1490	$9·793 \times 10^{-1}$	$−0·019 \times 10^{-1}$	1700	$6·790 \times 10^{-1}$	$−0·010 \times 10^{-1}$
1500	$9·604 \times 10^{-1}$	$−0·018 \times 10^{-1}$	1710	$6·687 \times 10^{-1}$	$−0·010 \times 10^{-1}$
1510	$9·421 \times 10^{-1}$	$−0·018 \times 10^{-1}$	1720	$6·585 \times 10^{-1}$	$−0·010 \times 10^{-1}$
1520	$9·243 \times 10^{-1}$	$−0·017 \times 10^{-1}$	1730	$6·487 \times 10^{-1}$	$−0·010 \times 10^{-1}$
1530	$9·070 \times 10^{-1}$	$−0·017 \times 10^{-1}$	1740	$6·391 \times 10^{-1}$	$−0·009 \times 10^{-1}$
1540	$8·903 \times 10^{-1}$	$−0·016 \times 10^{-1}$	1750	$6·297 \times 10^{-1}$	$−0·009 \times 10^{-1}$
1550	$8·741 \times 10^{-1}$	$−0·016 \times 10^{-1}$	1760	$6·206 \times 10^{-1}$	$−0·009 \times 10^{-1}$
1560	$8·583 \times 10^{-1}$	$−0·015 \times 10^{-1}$	1770	$6·117 \times 10^{-1}$	$−0·009 \times 10^{-1}$
1570	$8·430 \times 10^{-1}$	$−0·015 \times 10^{-1}$	1780	$6·030 \times 10^{-1}$	$−0·008 \times 10^{-1}$
1580	$8·282 \times 10^{-1}$	$−0·014 \times 10^{-1}$	1790	$5·946 \times 10^{-1}$	$−0·008 \times 10^{-1}$

TABLE 6

METHANATION REACTION

Gas phase equilibrium constants

(i) $CO + 3H_2 \rightleftharpoons CH_4 + H_2O$

$$K_p = \frac{P_{CH_4} \, P_{H_2O}}{P_{CO} \, P_{H_2}^3}$$

Values of K_p can be obtained from reciprocals of tables 4a and 4b.

(ii) $CO_2 + 4H_2 \rightleftharpoons CH_4 + 2H_2O$

$$K_p = \frac{P_{CH_4} \, P_{H_2O}^2}{P_{CO_2} \, P_{H_2}^4}$$

Table 6a

Temperature (°C)	$K_p = \dfrac{P_{CH_4} \, P_{H_2O}^2}{P_{CO_2} \, P_{H_2}^4}$
200	9.509×10^8
250	1.377×10^7
300	3.998×10^5
350	1.980×10^4
400	1.491×10^3
450	1.570×10^2
500	2.171×10^1
550	3.761
600	7.868×10^{-1}
650	1.930×10^{-1}
700	5.424×10^{-2}
750	1.714×10^{-2}
800	5.995×10^{-3}
850	2.292×10^{-3}
900	9.478×10^{-4}

Table 6b

Temperature (°F)	$K_p = \dfrac{P_{CH_4} \, P_{H_2O}^2}{P_{CO_2} \, P_{H_2}^4}$
400	6.313×10^8
500	6.457×10^6
600	1.492×10^5
700	6.349×10^3
800	4.324×10^2
900	4.266×10^1
1000	5.668
1100	9.600×10^{-1}
1200	1.988×10^{-1}
1300	4.874×10^{-2}
1400	1.378×10^{-2}
1500	4.405×10^{-3}
1600	1.565×10^{-3}
1700	6.092×10^{-4}
1800	2.570×10^{-4}

TABLE 7

AMMONIA SYNTHESIS

Gas phase equilibrium constants

$$N_2 + 3H_2 \rightleftharpoons 2NH_3$$

Table 7a

Temperature (°F)	\multicolumn{6}{c}{Mole percentage of ammonia at equilibrium}					
	650	750	850	950	1000
Pressure atm					
0	0	0	0	0	0
100	41	25	15·5	9·5	7·5
200	55	39	26·5	17	13·5
300	63·5	48	35·5	25	20
400		55	42	30	25

Table 7b

Temperature (°C)	\multicolumn{6}{c}{Mole percentage of ammonia at equilibrium}					
	350	400	450	500	550
Pressure atm					
0	0	0	0	0	0
100	38	25	16·5	10	7
200	52·5	39	27	18·5	12·5
300	61·5	48	36	27	18
400		55	43	32	23

TABLE 8

SPECIFIC HEATS OF CATALYSTS

The specific heat of a catalyst may be estimated as the sum of the contributions from each of its constituents. When the catalyst consists only of metal oxides the calculation is straightforward, but many catalysts contain residual carbonates or water which must be taken into account. A calculation of the heat requirement to raise the temperature of a converter requires information on the normal increase in specific heat with temperature and also the heat required to drive off water or carbon dioxide, or the change in heat content due to chemical reactions such as reduction; there will be a decrease in both the weight and the specific heat once the catalyst has lost its water. Similarly the specific heat of a catalyst in the reduced state will usually be slightly different from that in the oxidised state.

Fortunately, it is rarely necessary to know the specific heat of a catalyst accurately, and approximate values are normally adequate.

Table 8 lists approximate values for the specific heat of typical catalysts used on ammonia plants; the values have been calculated for catalysts, in the manufactured state, under loss free conditions (i.e. after water and CO_2 have been driven off).

Specific heats (values are given in cal/g °C and also in Btu/lb °F)

Temperature (°C)	20	100	200	300	400	500	800
Temperature (°F)	68	212	392	572	752	932	1472
Zinc oxide (ICI catalyst 32–4)	0·12	0·13	0·14	0·15	0·15		
Cobalt molybdate (ICI catalyst 41–3/4)	0·19	0·22	0·24	0·26	0·27	0·27	
Primary (ICI catalyst 57–1/46–1) Secondary (ICI catalyst 54–2) Methanation (ICI catalyst 11–3)	0·18	0·20	0·21	0·22	0·23	0·24	0·25
HT shift (ICI catalyst 15–4/5)	0·16	0·17	0·19	0·20	0·22	0·23	
LT shift (ICI catalyst 52–1)	0·16	0·16	0·17	0·17			
Ammonia synthesis (ICI catalyst 35–4)	0·15	0·17	0·19	0·21	0·23	0·25	

Approximate specific heats for ammonia plant catalysts, calculated for loss free catalyst

C_p in cal/g °C or Btu/lb °F

TABLE 9

SPECIFIC HEATS OF PURE GASES

Table 9a

kcal/Nm^3 °C											
Temperature (°C)	20	100	200	300	400	500	600	700	800	900	1000
Hydrogen	0·31	0·31	0·31	0·31	0·31	0·31	0·32	0·32	0·33	0·33	0·33
Nitrogen	0·31	0·31	0·32	0·32	0·33	0·33	0·34	0·34	0·35	0·36	0·36
Methane	0·38	0·42	0·48	0·54	0·59	0·64	0·71	0·77	0·79	0·82	0·86
Carbon monoxide	0·31	0·31	0·32	0·32	0·33	0·33	0·34	0·34	0·35	0·36	0·36
Carbon dioxide	0·41	0·43	0·47	0·50	0·54	0·55	0·56	0·57	0·58	0·59	0·60
Water vapour	0·36	0·36	0·37	0·37	0·37	0·37	0·37	0·37	0·37	0·38	0·38
Ammonia	0·38	0·41	0·43	0·46	0·49	0·53	0·57	0·59	0·61	0·62	0·65
Air	0·32	0·32	0·33	0·33	0·33	0·34	0·34	0·35	0·35	0·35	0·36

Table 9b

Btu/lb mole °F										
Temperature (°F)	60	200	400	600	800	1000	1200	1400	1600	1800
Hydrogen	7·0	7·0	7·0	7·0	7·0	7·1	7·2	7·3	7·4	7·5
Nitrogen	6·9	7·0	7·1	7·3	7·4	7·5	7·7	7·8	7·9	8·1
Methane	8·4	9·4	10·8	12·2	13·7	15·2	16·4	17·5	18·3	19·1
Carbon monoxide	6·9	7·0	7·1	7·3	7·4	7·5	7·7	7·8	7·9	8·1
Carbon dioxide	9·1	9·7	10·6	11·4	12·0	12·4	12·7	13·0	13·3	13·5
Water vapour	8·1	8·1	8·2	8·3	8·3	8·3	8·3	8·3	8·5	8·5
Ammonia	8·4	9·0	9·7	10·3	11·4	12·3	13·0	13·5	13·9	14·4
Air	7·2	7·2	7·3	7·4	7·5	7·6	7·7	7·8	7·9	8·1

TABLE 10

SPECIFIC HEATS OF PROCESS GAS MIXTURES

The composition of the process gases used for ammonia production do not vary greatly from plant to plant. The specific heat of the mixture is little affected by small changes in composition. Consequently tables 10a and 10b, which are derived for the indicated gas compositions, have more general application whenever approximate values of specific heats are required.

Table 10a

C_p (kcal/N m^3 °C)

Location	Composition								Specific heat at T°C									
	N_2	A	H_2	CO	CO_2	CH_4	NH_3	H_2O	20	100	200	300	400	500	600	700	800	900	1000
Primary reformer exit	0·2	0	34·3	6·4	8·3	5·0	—	45·8	0·35	0·35	0·36	0·37	0·37	0·38	0·38	0·39	0·39	0·40	0·40
Secondary reformer exit	12·7	0·2	31·5	8·6	6·5	0·2	—	40·3	0·34	0·34	0·35	0·35	0·35	0·36	0·36	0·36	0·37	0·37	0·38
HT shift exit	12·7	0·2	37·9	2·2	12·9	0·2	—	33·9	0·34	0·35	0·35	0·36	0·36	0·37	0·37	0·37	0·38	0·38	0·39
LT shift exit	12·7	0·2	39·8	0·3	14·8	0·2	—	32·0	0·34	0·35	0·36	0·36	0·37	0·37	0·37	0·38	0·39	0·39	0·39
CO_2 absorber exit	23·6	0·3	73·9	0·6	0·3	0·3	—	1·0	0·31	0·31	0·32	0·32	0·32	0·32	0·33	0·33	0·34	0·34	0·34
Methanator exit	24·0	0·3	72·1	—	—	1·3	—	2·3	0·31	0·31	0·32	0·32	0·32	0·32	0·33	0·33	0·34	0·34	0·35
Synthesis loop feed	24·6	0·3	73·8	—	—	1·3	—	—	0·31	0·31	0·32	0·32	0·32	0·32	0·33	0·33	0·34	0·34	0·35
Recycle gas	19·6	3·2	59·0	—	—	12·8	5·4	—	0·32	0·33	0·34	0·35	0·36	0·37	0·39	0·40	0·41	0·41	0·42

Table 10b

C_p (Btu/lb mole °F)

Location	Composition								Specific heat at T °F									
	N_2	A	H_2	CO	CO_2	CH_4	NH_3	H_2O	60	200	400	600	800	1000	1200	1400	1600	1800
Primary reformer exit	0·2	0	34·3	6·4	8·3	5·0	—	45·8	7·8	7·8	8·1	8·2	8·4	8·5	8·7	8·8	8·9	9·0
Secondary reformer exit	12·7	0·2	31·5	8·6	6·5	0·2	—	40·3	7·6	7·7	7·8	7·9	8·0	8·0	8·1	8·2	8·3	8·4
HT shift exit	12·7	0·2	37·9	2·2	12·9	0·2	—	33·9	7·7	7·8	7·9	8·1	8·2	8·2	8·4	8·5	8·6	8·7
LT shift exit	12·7	0·2	39·8	0·3	14·8	0·2	—	32·0	7·7	7·8	8·0	8·1	8·2	8·3	8·5	8·6	8·7	8·8
CO_2 absorber exit	23·6	0·3	73·9	0·6	0·3	0·3	—	1·0	7·0	7·0	7·1	7·1	7·1	7·2	7·3	7·5	7·6	7·7
Methanator exit	24·0	0·3	72·1	—	—	1·3	—	2·3	7·0	7·0	7·1	7·1	7·2	7·2	7·4	7·5	7·6	7·8
Synthesis loop feed	24·6	0·3	73·8	—	—	1·3	—	—	7·0	7·0	7·1	7·1	7·1	7·2	7·4	7·5	7·6	7·8
Recycle gas	19·6	3·2	59·0	—	—	12·8	5·4	—	7·2	7·4	7·7	7·9	8·2	8·5	8·8	9·0	9·2	9·5

TABLE 11

HEAT AND FREE ENERGY OF FORMATION

(i) Gases

	$\Delta H_{f(25°C)}$ (kcal/g mole)	$\Delta F_{f(25°C)}$ (kcal/g mole)
Naphtha assumed to be n-heptane (l)	−53.6	+0.42
CH_4 (g)	−17.889	−12.14
CO	−26.416	−32.808
CO_2	−94.052	−94.260
H_2O (g)	−57.7979	−54.6351
NH_3 (g)	−10.96	−3.903

(ii) Solids

	$\Delta H_{f(25°C)}$ (kcal/g mole)	$\Delta F_{f(25°C)}$ (kcal/g mole)
CoO	−57.5	
CoS	−22.3	−19.8
CrO_3	−139.3	
Cr_2O_3	−268.8	−249.3
Cu	0	0
Cu_2O	−43.0	−38.13
CuO	−38.5	−31.9
Fe	0	0
Fe_3O_4	−266.9	−242.3
Fe_2O_3	−198.5	−179.1
Ni	0	0
NiO	−58.4	−51.7

TABLE 12

HEAT RELEASED DURING CATALYST REDUCTION

Catalyst	ICI No.	$\Delta H_{R(25°C)}$	T (°C)	$\Delta H_{R(T)}$
Primary reforming	46–1	+1·69	750	−6·3
	57–1	+2·57	750	−9·5
Secondary reforming	54–2	+1·45	750	−5·4
High-temperature shift	15–4/5	+7·3 (Fe_2O_3) −8·3 per 1% CrO_3	400	+12·15
Low-temperature shift	52–1	−82·5	200	−85·5
Methanation	11–3	+1·7	400	−2·55
Ammonia synthesis	35–4	+139	400	+127

Column 3 gives the heat of reduction at 25°C, in kcal/kg of catalyst.
Column 5 gives the heat of reduction at the temperature indicated in column 4, again in kcal/kg of catalyst.

TABLE 13

HEAT RELEASED DURING CATALYST OXIDATION

Heats are calculated per kilogramme of catalyst in the form it is charged to the converter.

$\Delta H_{R(25°C)}$

Carbon burn-off
$C + \frac{1}{2}O_2 \rightarrow CO$ To remove 1 per cent C from 1 kg charged catalyst -11 kcal
$C + O_2 \rightarrow CO_2$ To remove 1 per cent C from 1 kg charged catalyst -78.4 kcal

Cobalt molybdate
$CoS + \frac{3}{2}O_2 \rightarrow CoO + SO_2$ To remove 1 per cent S from 1 kg charged catalyst -33.2 kcal

Primary and secondary reforming catalysts
$Ni + H_2O \rightarrow NiO + H_2$ oxidation with steam (and reduction)   negligible
For oxidation without steam: *see* methanation.

HT shift
$2Fe_3O_4 + \frac{1}{2}O_2 \rightarrow 3Fe_2O_3$ To oxidise 1 kg charged catalyst -343 kcal

LT shift
$Cu + \frac{1}{2}O_2 \rightarrow CuO$ To oxidise 1 kg charged catalyst -165 kcal

Methanation
$Ni + \frac{1}{2}O_2 \rightarrow NiO$ To oxidise 1 kg charged catalyst -164 kcal

Ammonia synthesis
$3Fe + 2O_2 \rightarrow Fe_3O_4$ To oxidise 1 kg charged catalyst -925 kcal

Table 14 *overleaf*

TABLE 14

PRESSURE DROP THROUGH CONVERTERS

Pressure drop data for fixed beds have been correlated by Carmen [*Trans. Inst. Chem. Eng.* **15**, 150 (1937)], where the symbols are defined. He plotted a pressure drop term against Reynolds number.

That is $\dfrac{\Delta P e^3}{h s_0 (1-e) \rho u^2}$ against $(Re) = \dfrac{u\rho}{s_0 \mu (1-e)}$

The correlation can be rearranged and expressed in more readily available units with the result that

$$\Delta P = B h^3 S^2 M T / P$$

where B = a quantity derived from a table by reference to the value of SMh/μ,
  S = space velocity (h^{-1}),
  M = average molecular weight of gas,
  h = height of bed (feet or metres),
  μ = viscosity of gas (lb ft^{-1} h^{-1} or centipoises),
  T = average temperature of gas in reactor (Rankine or Kelvin),
  P = average pressure in reactor [lb/in^2 (abs.) or kg cm^{-2} (abs.)].

ΔP is obtained in lb/in^2 or kg cm^{-2}. The factor B is dependent on SMh/μ.

To calculate a pressure drop, first calculate M/μ (see table 15), and evaluate SMh/μ, and then obtain B in ft lb h units from table 14a, or in metric units from table 14b. Substitution of the value of B in the equation given above gives the required pressure drop.

The values of pressure drop calculated from tables 14a and b are those expected in beds of packed catalyst. In use the bed packs down, causing the voidage in the bed to decrease and the pressure drop to increase to these expected values.

Some typical values of M/μ are given in table 15.

Pressure drop

Catalyst size Dia.×length, or dia.	3×3 mm	5.4×3.6 mm	5×5 mm	6×6 mm	8×8 mm	10×10 mm	8.5×11.3 mm	17×17 mm	17×17 mm	1/8–3/16 in	1/4–3/8 in
Shape	cylinder	cylinder	cylinder	cylinder	cylinder	cylinder	cylinder	ring	ring	spheres	granules
ICI Catalyst No.	—	11-3, 15-5, 41-3, 52-1	—	—	—	—	15-4	46-1, 57-1 Tubular reformer	54-2	32-4	35-4
SMh/μ (ft lb h)											
2.0×10^6	9.0×10^{-13}	4.4×10^{-13}	7.5×10^{-13}	5.2×10^{-13}	3.2×10^{-13}	2.2×10^{-13}	1.4×10^{-13}	9.4×10^{-14}	5.6×10^{-14}	4.4×10^{-13}	4.0×10^{-13}
3.0×10^6	5.6×10^{-13}	3.1×10^{-13}	5.4×10^{-13}	4.2×10^{-13}	2.5×10^{-13}	1.8×10^{-13}	1.2×10^{-13}	7.9×10^{-14}	4.7×10^{-14}	3.1×10^{-13}	3.3×10^{-13}
6.0×10^6	4.0×10^{-13}	2.5×10^{-13}	3.7×10^{-13}	2.8×10^{-13}	2.0×10^{-13}	1.5×10^{-13}	1.0×10^{-13}	6.4×10^{-14}	3.8×10^{-14}	2.5×10^{-13}	2.5×10^{-13}
1.0×10^7	3.4×10^{-13}	2.2×10^{-13}	3.2×10^{-13}	2.5×10^{-13}	1.8×10^{-13}	1.4×10^{-13}	9.6×10^{-14}	5.6×10^{-14}	3.4×10^{-14}	2.2×10^{-13}	2.1×10^{-13}
3.0×10^7	2.6×10^{-13}	1.9×10^{-13}	2.6×10^{-13}	2.1×10^{-13}	1.5×10^{-13}	1.2×10^{-13}	8.6×10^{-14}	4.7×10^{-14}	2.9×10^{-14}	1.9×10^{-13}	1.5×10^{-13}
SMh/μ (metric units)											
1.5×10^6	2.2×10^{-12}	1.1×10^{-12}	1.8×10^{-12}	1.3×10^{-12}	7.7×10^{-13}	5.4×10^{-13}	3.6×10^{-13}	2.4×10^{-13}	1.4×10^{-13}	1.1×10^{-12}	1.0×10^{-12}
2.0×10^6	1.6×10^{-12}	8.2×10^{-13}	1.4×10^{-12}	1.1×10^{-12}	6.6×10^{-13}	4.8×10^{-13}	3.2×10^{-13}	2.0×10^{-13}	1.2×10^{-13}	8.2×10^{-13}	8.7×10^{-13}
4.0×10^6	1.0×10^{-12}	6.2×10^{-13}	9.7×10^{-13}	7.7×10^{-13}	5.1×10^{-13}	3.8×10^{-13}	2.6×10^{-13}	1.6×10^{-13}	9.7×10^{-14}	6.2×10^{-13}	6.5×10^{-13}
8.0×10^6	8.2×10^{-13}	5.4×10^{-13}	7.7×10^{-13}	5.9×10^{-13}	4.4×10^{-13}	3.4×10^{-13}	2.4×10^{-13}	1.4×10^{-13}	8.6×10^{-14}	5.4×10^{-13}	5.2×10^{-13}
2.0×10^7	2.6×10^{-13}	4.8×10^{-13}	6.5×10^{-13}	5.2×10^{-13}	3.8×10^{-13}	2.9×10^{-13}	2.2×10^{-13}	1.2×10^{-13}	7.3×10^{-14}	4.8×10^{-13}	3.7×10^{-13}

Values of B in ft lb h units / Values of B in metric units

TABLE 15

MOLECULAR WEIGHT/VISCOSITY RATIOS

Some values of molecular weight/viscosity are given for typical gas streams. For most duties the accuracy of these values is adequate for the evaluation of pressure drop, in conjunction with tables 14a and 14b.

Duty	M	M/μ (ft lb h units)	M/μ (metric units)
Primary reformer	15	190	460
Secondary reformer	16	150	360
HT CO shift	16	290	700
LT CO shift	16	370	890
Methanation (NH_3 plant)	8·6	200	480
Methanation (H_2 plant)	2·4	70	160

EXAMPLES OF CALCULATIONS

TABLE 16

EXAMPLES OF CALCULATIONS

EXAMPLE 1

Sulphur absorption

Calculation of partial pressure of hydrogen sulphide over partially sulphided zinc oxide

Calculate the partial pressure at 410°C, in a gas stream containing 1 per cent H_2O, in a total pressure of 30 atm.

At 410°C, from table 3a $K_p = 5.57 \times 10^6$.

$$\frac{P_{H_2O}}{P_{H_2S}} = 5.57 \times 10^6$$

$$P_{H_2S} = \frac{0.3}{5.57 \times 10^6} = 5.4 \times 10^{-8} \text{ atm}$$

H_2S concentration $= 1.8 \times 10^{-3}$ ppm.

Similarly at 410°C, and 30 atm, in a gas stream containing 50 per cent steam

$$K_p = \frac{P_{H_2O}}{P_{H_2S}} = 5.57 \times 10^6$$

$$P_{H_2S} = \frac{15}{5.57 \times 10^6} = 2.7 \times 10^{-6} \text{ atm}$$

H_2S concentration $= 0.09$ ppm.

Under low-temperature shift guard conditions at 230°C, 30 atm, and with 33 per cent steam.

$$K_p = 1.19 \times 10^8$$

$$P_{H_2S} = \frac{10}{1.19 \times 10^8} = 8.4 \times 10^{-8} \text{ atm}$$

H_2S concentration $= 2.8 \times 10^{-3}$ ppm.

EXAMPLE 2

Sulphur absorption

Calculation of the average sulphur concentration in a bed of zinc oxide catalyst

Average ingoing sulphur concentration 35 ppm v/v
Gas rate 4600 N m^3/h.
Catalyst volume 11·5 m^3.
Duration of run 1½ years.
Weight of sulphur per annum is

$$\text{v/v ppm} \times 10^{-6} \times \text{space velocity} \times 24 \times 365 \times 32/22\cdot 4 \text{ kg,}$$

i.e. v/v ppm × (S.V.) × $1\cdot 25 \times 10^{-2}$ kg S/m^3 catalyst year.

In 1½ years there are

$$1\cdot 5 \times 35 \times \frac{4600}{11\cdot 5} \times 1\cdot 25 \times 10^{-2} = 262\cdot 5 \text{ kg S/m}^3 \text{ catalyst.}$$

But   (65·37 + 16·00) kg ZnO + 32·06 kg S → (65·37 + 32·06) kg ZnS

i.e.   1 kg ZnO + x kg S → (1 + ½x) kg ZnS.

However, 1 m^3 fresh catalyst 32–4 is 1100 kg.
Hence   1 m^3 fresh catalyst 32–4 becomes {1100 + (262·5/2)} = 1231·25 kg used catalyst.
Therefore the average sulphur content of the used catalyst is

$$\frac{262\cdot 5}{1231\cdot 25} \times 100 = 21\cdot 3 \text{ per cent w/w S}$$

EXAMPLE 3

Steam reformer calculations

(a) *Graphical solutions*

The exit gas of a reformer, operating under known conditions can readily be estimated, to a fair degree of accuracy, by use of the graphs in Chapter 5 (figs. 28–30 for methane reforming, figs. 31–34 for reforming other hydrocarbons). Thus,

$$\begin{aligned}
\text{inlet gas} &= \text{methane} + \text{steam}, \\
\text{steam ratio} &= 3\cdot9, \\
\text{exit pressure} &= 32\cdot0 \text{ atm (absolute)}, \\
\text{exit temperature} &= 808°C.
\end{aligned}$$

If the exit gas is at equilibrium its composition can be found from figs. 28–30.

$$32\cdot0 \text{ atm} = 455\cdot7 \text{ lb/in}^2 \text{ (gauge)}$$

The correct pressure is found on the pressure scale at the left hand of the lower graph (fig. 28). A horizontal line at this pressure is drawn to meet a curve for 808°C. A vertical line is drawn from this point on the lower curve to join an upper curve for steam ratio 3·9. A horizontal line from this point on the upper curve to the right-hand margin permits the equilibrium methane concentration (per cent) to be read from the scale. A similar technique, with figures 29 and 30, permits CO and CO_2 to be estimated. The hydrogen can then be calculated by difference: $(100 - CH_4 - CO - CO_2)$.

Values are: $CH_4 = 6$, $CO = 10$, $CO_2 = 10$, $H_2 = 74$.

The curves can be used, following a reverse line, to estimate the pressure at which a given composition will be in equilibrium. Alternatively the temperature can be estimated when composition, pressure and steam ratio are known (or steam ratio found with the other terms being known).

Under normal running conditions the methane content of the exit gas does not reach equilibrium, but the water-gas shift reaction is at equilibrium. It is not, strictly, possible to use the curves under these circumstances. However, the water-gas equilibrium does not change rapidly with temperature. An approximate answer can therefore be obtained by using the temperature at which the methane would be at equilibrium.

For example, in the previous calculation, it may be known that the approach to equilibrium was 16°C. The methane steam reaction should be at equilibrium at $808 - 16 = 792°C$. The exit gas has an approximate composition obtained by use of curves for 792°C in figs. 28–30.

Higher accuracy is required for some purposes. This can readily be obtained by use of a graphical solution as the starting point for the computation described in the following example.

APPENDICES

(b) *Computational solutions*

Assuming a reformer is operating under known conditions the exit gas composition can be calculated. For example:

Inlet gas = methane + steam
Steam ratio = 3·9
Exit temperature = 808°C
Exit pressure = 32·0 atm (absolute)
Approach to equilibrium = 16°C

The exit gas composition can be evaluated by the technique described on p.94. This involves the solution of three simultaneous equations. One equation is a mass balance, and the other equations are the equilibria of the methane steam and water-gas shift reactions. An iterative technique to find the solution is set out. It starts with any solution of the mass balance equation and corrects this solution in a series of steps until the correct solution is found. When no information is available about the final ratio of CO to CO_2 it is advantageous to initially assume it is 1·0, and also that the conversion of methane (which can lie between 0 and 100 per cent) is 50 per cent. Hence the following equation:

$$CH_4 + 3·9H_2O \rightarrow 0·25CO + 0·25CO_2 + 0·5CH_4 + 3·15H_2O + 1·75H_2 \quad (1)$$

K_{ms} (at $808 - 16°C = 792°C$) = $1·294 \times 10^2 + 2 \times 0·033 \times 10^2 = 136·0$

But $\dfrac{K_{ms}}{P^2} = \dfrac{136·0}{32^2} = 0·13281$

$$= \frac{(CO)(H_2)^3}{(CH_4)(H_2O)} \times \frac{1}{\{(CO)+(CO_2)+(CH_4)+(H_2O)+(H_2)\}^2}$$

Let $(CO)+(CO_2)+(CH_4)+(H_2O)+(H_2) = V$

then $\qquad (CH_4) = \dfrac{(CO)(H_2)^3}{0·13281 \, (H_2O)V^2} \qquad (2)$

And $\qquad K_{wgs}$ at $808°C = 0·9968 - 3 \times 0·0035 = 0·9863$

As $\qquad K_{wgs} = \dfrac{(CO_2)(H_2)}{(CO)(H_2O)} = 0·9863$

Then $\qquad (CO) = \dfrac{(CO_2)(H_2)}{0·9863 \, (H_2O)} \qquad (3)$

For convenience the mass balance, equation (1), can be written

$CH_4 = 0·50 \quad CO = 0·25 \quad CO_2 = 0·25 \quad H_2O = 3·15 \quad H_2 = 1·75$

[continued overleaf]

From equation (2)

$$(CH_4) = \frac{0.25 \times 1.75^3}{0.13281 \times 3.15 \times 5.90^2} = 0.09; \text{ change in CO} = 0.50 - 0.09 = 0.41$$

This indicates that the exit methane is below our initial assumed value. It can be shown, however, that the value 0·09 is less accurate than the initial value. Experience shows that the values of CH_4 must be adjusted by 40 per cent of the calculated amount if a rapid approach to the solution is required.

Since $\Delta CH_4 = -0.41$; $0.4\, CH_4 = -0.16 = \Delta_2$, the mass balance becomes

$$CH_4 + \Delta_2,\ CO - \Delta_2,\ CO_2 \text{ no change},\ H_2O + \Delta_2,\ H_2 - 3\Delta_2,$$

i.e. $\quad CH_4 = 0.34 \quad CO = 0.41 \quad CO_2 = 0.25 \quad H_2O = 2.99 \quad H_2 = 2.23$

From equation (3)

$$(CO) = \frac{0.25 \times 2.23}{0.9863 \times 2.99} = 0.19; \text{ change in CO} = 0.41 - 0.19 = 0.22$$

Again it is necessary to adjust the CO by 40 per cent of the calculated amount
Since $\Delta CO = -0.22$; $0.4\, \Delta CO = -0.09 = \Delta_3$, the mass balance becomes

$$CH_4 \text{ no change},\ CO + \Delta_3,\ CO_2 - \Delta_3,\ H_2O + \Delta_3,\ H_2 - \Delta_3,$$

i.e. $\quad CH_4 = 0.34 \quad CO = 0.32 \quad CO_2 = 0.34 \quad H_2O = 2.90 \quad H_2 = 2.32$

Similarly the V in equation (2) can be adjusted

$$\text{new } V = \text{old } V - 2\Delta_2 = 5.90 + 0.32 = 6.22$$

This solution is more correct than the original assumption.
The process is therefore continued.

$$\Delta CH_4 = -0.072$$

$$(CH_4) = \frac{0.32 \times 2.32^3}{0.13281 \times 2.90 \times 6.22^2} = 0.268$$

$$0.4 \Delta CH_4 = -0.029 = \Delta_2$$

$$CH_4 = 0.311 \quad CO = 0.349 \quad CO_2 = 0.340 \quad H_2O = 2.861 \quad H_2 = 2.417$$

$$\Delta CO = 0.059$$

$$(CO) = \frac{0.340 \times 2.417}{0.9863 \times 2.861} = 2.90$$

$$0.4 \Delta CO = 0.024 = \Delta_3$$

$$CH_4 = 0.311 \quad CO = 0.325 \quad CO_2 = 0.364 \quad H_2O = 2.837 \quad H_2 = 2.441$$

V has become 6·278.

At this stage the answer is reaching the limits of the slide-rule; each figure is within 2 per cent of its correct solution.

If further accuracy is required the iteration can be continued, after the mass balance (equation 1) has been checked.

APPENDICES

$$\Delta CH_4 = 0.009$$

(Equation 2) $CH_4 = \dfrac{0.325 \times (2.441)^3}{0.13281 \times 2.837 \times 6.278^2} = 0.320$

$$0.4\Delta CH_4 = 0.0036 = \Delta_2$$

$CH_4 = 0.3146 \quad CO = 0.3214 \quad CO_2 = 0.3640 \quad H_2O = 2.8406 \quad H_2 = 2.4302$

$$\Delta CO = -0.0043$$

(Equation 3) $CO = \dfrac{0.364 \times 2.4302}{0.9863 \times 2.8406} = 0.3157$

$$0.4\Delta CO = -0.0017 = \Delta_3$$

$CH_4 = 0.3146 \quad CO = 0.3197 \quad CO_2 = 0.3657 \quad H_2O = 2.8389 \quad H_2 = 2.4319$

$$\Delta CH_4 = -0.0035$$

(Equation 2) $CH_4 = \dfrac{0.3197 \times (2.4319)^3}{0.13281 \times 2.8389 \times (6.2708)^2} = 0.3101$

$$0.4\Delta CH_4 = -0.0014 = \Delta_2$$

$CH_4 = 0.3132 \quad CO = 0.3211 \quad CO_2 = 0.3657 \quad H_2O = 2.8375 \quad H_2 = 2.4361$

$$\Delta CO = -0.0028$$

(Equation 3) $CO = \dfrac{0.3657 \times 2.4361}{0.9863 \times 2.8375} = 0.3183$

$$0.4\Delta CO = -0.0011 = \Delta_3$$

$CH_4 = 0.3132 \quad CO = 0.3200 \quad CO_2 = 0.3668 \quad H_2O = 2.8364 \quad H_2 = 2.4372$

$$\Delta CH_4 = -0.00074$$

(Equation 2) $CH_4 = \dfrac{0.3200 \times (2.4372)^2}{0.13281 \times 2.8364 \times (6.2736)^2} = 0.31246$

$$0.4\Delta CH_4 = -0.00022 = \Delta_2$$

$CH_4 = 0.31298 \quad CO = 0.32022 \quad CO_2 = 0.36680 \quad H_2O = 2.83566 \quad H_2 = 2.43786$

$$\Delta CO = -0.00050$$

(Equation 3) $CO = \dfrac{0.36680 \times 2.43786}{0.9863 \times 2.83566} = 0.31972$

$$0.4\Delta CO = -0.00020 = \Delta_3$$

$CH_4 = 0.31298 \quad CO = 0.32002 \quad CO_2 = 0.36700 \quad H_2O = 2.83546 \quad H_2 = 2.43806$

Hence exit gas has following composition (per cent)

CH_4	9·10
CO	9·31
CO_2	10·67
H_2	70·92
Total 'dry'	100·00
H_2O	82·47

Approach to equilibrium (computation)

When the exit gas analysis, including steam, is known accurately (or inlet steam ratio and dry exit gas analysis), together with the pressure, the approach to equilibrium can be calculated.

For example pressure 13·5 atm (absolute)

$$\begin{array}{ll} CH_4 & 5\cdot98 \\ CO & 9\cdot88 \\ CO_2 & 10\cdot12 \\ H_2 & 70\cdot12 \\ N_2 & 3\cdot90 \\ \text{Total} & 100\cdot00 \\ H_2O & 60\cdot81 \\ \text{Total} & 160\cdot81 \end{array}$$

$$K_{wgs} = \frac{10\cdot12 \times 70\cdot12}{9\cdot88 \times 60\cdot81} = 1\cdot181 \quad T_{eq} = 760°C$$

$$K_{ms} = \frac{9\cdot88 \times 70\cdot12^3}{5\cdot98 \times 60\cdot81} \times \left(\frac{12\cdot5}{160\cdot81}\right)^2 = 56\cdot60 \quad T_{eq} = 757°C$$

Approach = 3°C

The approach can be calculated from other data by a similar technique.

EXAMPLE 4

(a) High-temperature shift

Assuming the gas composition exit the secondary reformer and the HT shift (given in table 10a) and an exit temperature of 450°C what is the equilibrium CO concentration, and what is the approach to equilibrium?

$$K = \frac{P_{CO_2} \cdot P_{H_2}}{P_{CO} \cdot P_{H_2O}} = \frac{12 \cdot 9 \times 37 \cdot 9}{2 \cdot 2 \times 33 \cdot 9} = 6 \cdot 55$$

From table 4a

$$K = K_p = 6 \cdot 55 \quad \text{when} \quad T = 463°C$$

Therefore equilibrium approach is 13°C.

The adiabatic equilibrium CO concentration may be calculated assuming an average temperature rise of $\sim 10°C$ per 1 per cent of CO converted.

CO	H$_2$O	CO$_2$	H$_2$	K	T for K = K$_p$	Actual T
2·2	33·9	12·9	37·9	6·55	463	450
2·15	33·85	12·95	37·95	6·75	459	450·5
2·1	33·8	13·0	38·0	7·3	450	451

From this it can be seen that the adiabatic equilibrium CO concentration is 2·11 per cent (wet).

(b) Low-temperature shift

Assuming the gas composition exit the HT and LT shift converters shown in table 10a, and an exit temperature of 250°C, what is the equilibrium CO concentration, and what is the approach to equilibrium?

$$K = \frac{P_{CO_2} \cdot P_{H_2}}{P_{CO} \cdot P_{H_2O}} = \frac{14 \cdot 8 \times 39 \cdot 8}{0 \cdot 3 \times 32} = 61 \cdot 3$$

$$K = K_p = 61 \cdot 3 \text{ at } 271°C$$

Therefore approach to equilibrium = 21°C.

The conversion of a further proportion of the CO will not significantly affect the exit temperature.

at 250°C $K_p = 86 \cdot 5$

CO	H$_2$O	CO$_2$	H$_2$	K
0·3	32	14·8	39·8	61·3
0·25	31·95	14·85	39·85	74
0·23	31·93	14·87	39·87	80·7
0·22	31·92	14·88	39·88	84·5
0·21	31·91	14·89	39·89	94·5

Adiabatic equilibrium CO concentration = 0·22 per cent (wet)
Adiabatic equilibrium temperature = 250·8°C

EXAMPLE 5

Heat release during reduction

LT shift

It is assumed that at any point in the bed, catalyst and gas are at the same temperature.

Catalyst volume = 30 m³ (27 Te)   Gas flow = 18 000 N m³ h⁻¹
Space velocity = 600 h⁻¹
Inlet temperature = 200°C

(a) Reduction gas is 1 per cent hydrogen in nitrogen.

$$\text{Hydrogen supplied} = 6 \text{ N m}^3 \text{ h}^{-1}/\text{m}^3 \text{ catalyst}$$
$$= 180 \text{ N m}^3 \text{ h}^{-1}$$
$$= 8 \cdot 025 \text{ kg moles h}^{-1}$$

This will reduce 8·025 kg moles of CuO

$$= 638 \text{ kg of CuO}$$

The catalyst contains 34 per cent CuO

638 kg of CuO are contained in 1880 kg of catalyst

The heat release per hour (see table 12)

$$= 85 \cdot 5 \times 1880$$
$$= 1 \cdot 61 \times 10^5 \text{ kcal}$$

If temperature rise is ΔT °C, using specific heats from tables 8 and 12

$$(18\,000 \times 0 \cdot 32 \times \Delta T) + (1880 \times 0 \cdot 17 \times \Delta T) = 1 \cdot 61 \times 10^5$$
$$5760 \Delta T + 319 \Delta T = 1 \cdot 61 \times 10^5$$
$$\Delta T = 26 \cdot 5 °C$$

i.e. temperature rise is approximately 25°C.

(b) Reduction gas is 10 per cent hydrogen in nitrogen

$$\text{Hydrogen supplied} = 60 \text{ N m}^3 \text{ h}^{-1}/\text{m}^3 \text{ catalyst}$$
$$= 1800 \text{ N m}^3$$
$$\text{Heat released} = 1 \cdot 61 \times 10^6 \text{ kcal per hour}$$

Temperature rise $\approx 250°C$, which is clearly intolerable.

EXAMPLE 6

Oxidation of catalysts

(1) *HT shift*

(a) By air at 20°C at space velocity of 500 h^{-1} oxygen supplied per m^3 of catalyst.

$$= 105 \text{ m}^3 \text{ h}^{-1}$$
$$= 4\cdot47 \text{ kg moles/h}^{-1}$$

$2Fe_3O_4 + \frac{1}{2}O_2 \rightarrow 3Fe_2O_3$

Oxygen requirement for oxidation = $\frac{1}{6}$ mole per mole of Fe_2O_3 charged.

1 m^3 of catalyst = 1·3 Te
1·3 Te of catalyst contain 6·76 kg moles Fe_2O_3

Oxygen requirement = 6·76/6 = 1·13 kg moles/m^3.
Oxygen supply is therefore in excess of stoichiometric requirements.
Heat released by oxidation (see table 13) = 343 kcal/kg.

$$= 343 \times 1300 = 4\cdot46 \times 10^5 \text{ kcal.}$$

Using specific heats from tables 8 and 9

$$(500 \times 0\cdot33 \times \Delta T) + (1300 \times 0\cdot2 \times \Delta T) = 4\cdot46 \times 10^5$$
$$(165 \times 260)\Delta T = 4\cdot46 \times 10^5$$
$$\Delta T = 1050°C$$

The temperature rise obtained by admitting air to the reduced HT shift catalyst is enough to make the catalyst at least red-hot, and possibly also to fuse it.

(b) By nitrogen containing 1 per cent oxygen at a space velocity of 500 h^{-1}

Oxygen supplied per m^3 of catalyst = 5 m^3 h^{-1}
$$= 0\cdot223 \text{ kg moles h}^{-1}$$
Oxygen requirement for total oxidation = 1·13 kg moles/m^3
Therefore heat released per hour = 343 × 0·223/1·13 = 67·7 kcal/kg charged
$$= 0\cdot88 \times 10^5 \text{ kcal/m}^3$$
$$(500 \times 0\cdot33 \times \Delta T) + (1300 \times 0\cdot2 \times T) = 0\cdot88 \times 10^5$$
$$\Delta T \approx 200°C \text{ per hour}$$

This heating rate will be observed only while oxidation is incomplete. Under these conditions total oxidation would take about 5 hours.

(c) A full analysis of the problems associated with the oxidation (or reduction) of a catalyst can be calculated only if the rate of reaction is known. The next example has been estimated from the data used in example (b). It is quoted as a warning against the unjustified assumption that the temperature rise in a bed of catalyst cannot exceed the adiabatic temperature due to reaction.

Consider a bed of 10 m^3 of Fe_3O_4 catalyst, divided into ten equal sections. This is being oxidised by 10 000 Nm^3 h^{-1} of nitrogen containing 1 per cent oxygen. A kinetic assumption has been made that 50 per cent of the oxygen entering each section will be used in that section, until the catalyst is completely oxidised. (There is no information on which this assumption can be justified. This example demonstrates a hypothetical case.)

In 15 minutes the first section will be half oxidised, which will raise the temperature of gas and catalyst by 200°C. The hot gas leaving the first section will have heated the catalyst in the second section about 150°C, and reaction in this section will have raised the catalyst temperature a further 100°C, a total of 250°C. The process is continued down the converter.

In the next 15 minutes cold gas entering the first section will have been heated 50°C by the hot catalyst and a further 200°C by reaction. Again the process continues through the bed.

The overall pattern of temperatures through the bed at successive intervals of 15 minutes is shown on the facing page.

The adiabatic temperature rise from the complete reaction of 1 per cent oxygen with Fe_3O_4 is 407°C. Yet in 2 hours the peak temperature has risen by 730°C.

APPENDICES

Section of bed

Time (hours)	1	2	3	4	5	6	7	8	9	10
$\frac{1}{4}$		200	250	240	200	170	120	90	60	40	30
$\frac{1}{2}$		250	350	370	350	320	270	220	180	150	120
$\frac{3}{4}$		60	350	460	480	460	420	380	330	280	240
1		15	100	390	510	550	550	510	480	430	380
$1\frac{1}{4}$		4	30	110	400	540	590	600	575	540	500
$1\frac{1}{2}$		1	7	30	90	410	600	650	660	640	620
$1\frac{3}{4}$		—	2	9	29	120	440	640	710	720	710
2		—	—	2	9	37	138	470	630	700	730
$2\frac{1}{4}$		—	—	—	2	11	43	150	470	630	700
$2\frac{1}{2}$		—	—	—	—	3	13	46	151	471	630
$2\frac{3}{4}$		—	—	—	—	1	4	15	49	154	473
3		—	—	—	—	—	1	5	16	50	156

(2) *Ammonia synthesis catalyst*
By 1 per cent oxygen in nitrogen at a space velocity of 500 h^{-1}
Oxygen supplied per m^3 of catalyst = 0·223 kg mole h^{-1}.

$$3Fe + 2O_2 \rightarrow Fe_3O_4$$

Oxygen requirement = 2 kg moles per kg mole of Fe_3O_4 charged
1 m^3 catalyst = 2·75 Te (if bulk density = 2·75)
= 10·5 kg moles
Oxygen requirement = 21 kg moles.
The oxygen supply, 0·223 kg mole h^{-1}, limits the rate of heat release.
Heat released per hour
= 925 × 0·223/21 = 9·83 kcal/kg catalyst charged
= 0·270 × 10^5 kcal/m^3 catalyst charged
If temperature rise is $\Delta T°C$, using specific heats from tables 8 and 12
$(500 \times 0.33 \times \Delta T) + (2750 \times 0.21 \times \Delta T) = 0.270 \times 10^5$ kcal h^{-1}
$\Delta T \approx 40°C$ per hour

Subject Index

absorption of sulphur compounds, 49–55
acetylene:
 in feed gases, 104
 saturation, 3
acid washing plants, 47
activators, 29
activity, 5, 9, 147
 ammonia synthesis catalysts, 134–8
 catalyst testing, 32–3
 desulphurisation catalysts, 48
 high-temperature shift catalysts, 98
 insulators, 15
 loss of, 89, 103, 171, 173
 low-temperature shift catalysts, 110
 metals, 12–3
 methanation catalysts, 121
 methane reforming, 91
 reforming catalysts, 79, 81–2
 relationship to structure, 22
 requirements, 20–1
 semiconductors, 17
 zinc oxide, 50
adsorption, 12
air injection in secondary reformers, 79
alumina, 9, 26, 29, 110
 action of steam and water on, 28
 ammonia synthesis, 131, 134–5
 catalyst support, 18, 109
 cobalt molybdate, 49
 methanation catalysts, 121
 reforming catalysts, 79
 low-temperature shift catalysts, 110, 114
 reforming catalysts, 78, 80, 83
 secondary reformer catalyst protection, 90, 178
 spacers, 112
 spinel in methanation catalysts, 121
 stabiliser, 24, 82, 135
aluminium bonds, 132
alumino-silicates:
 ammonia synthesis, 135
 catalytic effects, 82–3
ammonia:
 converter programs, 155–60
 naphtha process, 64
 oxidation, 126
 specific heat, 128, 130
 synthesis, 3, 29, 65, 197
  catalysts, chapter 7, 172–4, 178–81;
   testing, 144–5

 kinetics, 141–3
 nitrogen for, 90
 pressure drop, beds of, 206–7
 reduction of, 172–4, 203
 poisons in, 86
 secondary reforming, 76–7
 studied by recirculatory reactor, 41–3
 thermodynamics, 126
ammonium carbamate, 117
amphoteric oxides, 82
area, surface, *see* surface area
Arrhenius equation, 2
arsenic:
 catalyst poisons, 86–8, 138–9
 deactivation of cobalt molybdate, 61
arsenious oxide, 86–7, 124
attrition testing, 44

B

Benfield process, 124
benzene:
 in naphtha feedstock, 88
 reforming, carbon formation in, 74
blanketing, 175–6, 179–80
butane:
 hydrocarbon feedstock, 46
 reforming:
  carbon formation in, 74
  cracking activities in, 82

C

calcium:
 aluminate bonding:
  methanation catalysts, 121
  primary reforming catalysts, 80, 82
 aluminosilicate crystals, 136–7
 bonds, 132
 oxide:
  ammonia synthesis catalysts, 131
  promotors, 135
  reforming catalysts, 78, 80–1
 stearate as die-wall lubricant, 25
carbon:
 dioxide:
  catalyst poisons, 138
  in hydrogenating gas, 61
  in methane reforming, 64–5, 68, 90
  in naphtha reforming, 66, 71, 92

SUBJECT INDEX

reforming equilibrium-gas composition, 77
removal process, poisons originating in, 86
disulphide, 104
formation, see hydrocarbon reforming, carbon formation and removal; high-temperature shift, carbon formation and methanation, carbon formation and removal
monoxide:
  catalyst poisons, 138
  effect on equilibrium gas-composition, 77
  hydrogenating gas, 61
  in high-temperature shift conversion, 98–109
  in low-temperature shift conversion, 109–117
  in methanation, 117–125
  in methane reforming, 65–7, 90
  in naphtha reforming, 66, 70, 92
  removal from feedstock, chapter 6
oxy-sulphide, 102, 104
catalysts:
  composite, 17–9
  disposal of, 180–1
  efficacy of, 8–10
  handling, chapter 9
  pressure drop, beds of, 107–9, 206–7
  reduction of, 168–74
  size, effect of, 143–5
  specific heats, 198
  supports, see supports
cement bonding, 79–80
ceramic oxide compound supports, 24
charging, 161–8
  catalyst testing, 32
  pellet strength requirements, 26–7
chemisorption:
  insulators, 12–3
  metals, 10–12
  semiconductors, 16
chlorine as catalyst poison, 28, 87, 113, 138–9
chrome, safety precautions with, 181
chromia stabiliser, 24
chromic oxide, 15
chromium sesquioxide, 98
coal, see hydropetrol
cobalt:
  catalysts:
    discharging, 176
    in hydrocarbon reforming, 78
  molybdate, 17, 48, 57
    carbon formation on, 62–3
    deactivation, 61
    in desulphurisation, 47–8, 53, 55
    in heptane desulphurisation, 58–61
    in high-temperature shift conversion, 98
    in methanation, 62
    redox reactions, 98
  oxide, 48
cold-shot converter, discharging, 179
computer programs, chapter 8
  RTC00, 151–3, 155

RTC01, 153
RTC02, 153
RTC03, 153
RTCP1, 155
RTK22, 157, 159
RTK25, 155–7
RTK26, 157–8
RU009, 160
continuous stirred tank reactors, 43
converters:
  ammonia cold-shot, 149, 155–7
  blanketing ammonia, 175
  costing, 148–9
  design of, 149–50
  multi-bed, 149–50
  pressure drop through, 206–7
  Simplex technique for design of, 149, 160
  tube-cooled, 150, 155, 157, 159–60
copper:
  catalyst poisons, 87
  catalysts:
    disposal of, 180
    for redox reaction, 97–8
    structural collapse of, 28
  compacts, 23
  crystal sizes, 21, 23
  hydrogenation-dehydrogenation activity of, 18, 109
  low-temperature shift catalysts, 109–110, 114
  oxide, 110
  surfaces, 21
cracking of hydrocarbons, 82
creosote, see hydropetrol
crushing strength tests, 44
crystal:
  growth, 23, 28, 79
  sizes, 22–4, 79
cyclones, 179

D

deactivation, 61, 111–3. See also poisoning
desulphurisation, chapter 4, 84–6, 189. See also sulphur compounds
desulphurisers, effect of voidage differences in, 37
deuterium, 91
diethanolamine, 124
diethyl sulphide, 59–61
differential reactors for catalyst testing, 38–9, 41–3
diffusion, 22. See also pore diffusion
di-isopropanolamine, 124
dimethyl disulphide, 59–61
dimethyl sulphide, 46
discharging, 105, 176, 178–80
disulphides:
  hydrogenolysis of, 58
  thermal decomposition of, 55–6
drum handling, 161–3

E

electron probe analysis, 135–7
electronic structure, 6
extrusion forming of catalysts, 29–30, 48

SUBJECT INDEX

F

ferric oxide, 101, 170
formaldehyde, 4
Friedel-Craft's catalysts, 13

G

gas:
  chemisorption, 112
  distribution, 36, 39
  -film diffusion limitation, 35
granulation formation of catalysts, 29–30
granule size, 143

H

halogen deactivation, 28, 111, 138–9
halogenation, 8
heptane:
  (n) in heptane desulphurisation, 59–61
  reforming, 84
high-temperature shift. *See also* water-gas shift
  calculations, 217
  catalysts, 97–100, 170–1
    carbon formation in, 101, 104–5, 174
    discharging, 176
    disposal of, 180–1
    pressure drop, beds of, 107, 206–7
    reactions with, 100, 105
    reduction of, 101, 170, 203
    regeneration, 174
    re-use of, 180
    stabilisation, 176–7
  converter design, 151–3
  equilibria, 192–5
hydrocarbon-reforming, chapter 5
  carbon formation and removal in, 73–6, 78, 86, 89, 96, 174–5
  catalysts, 77–83
    reduction of, 169
    regeneration of, 174–5
    re-use of, 180
  equilibria, 190–1
  heats of reaction, 72–3
  product gas composition, 65–72, 212–6
  supports, 78–81
hydrodesulphurisation, 57–61
  cobalt molybdate, 17
hydrofluoric acid treatment, 15
hydrogen:
  in naphtha reforming, 66, 72
  in desulphurisation, 57–60
  specific heat of, 128
  sulphide:
    in desulphurisation, 48, 53–5, 57, 62
    in high-temperature shift conversion, 102–5
    in hydrocarbon feedstocks, 46, 84
    in ICI catalyst 15-2, 99
    in low-temperature shift catalysts, 113
hydrogenolysis, 8
  organic sulphur compounds, 57–61

hydropetrol (hydrocarbon feedstock), 64

I

ICI 500 process, 64
ICI catalysts:
  general properties of, 187
  ICI catalyst 11–3; 18, 122, 172
  ICI catalyst 15–2; 98, 100, 103–4
  ICI catalyst 15–4/5; 18
  ICI catalyst 15–4; 100, 104, 153, 170, 174
    particle size, 106–7
    sulphate content, 102
  ICI catalyst 15–5; 100, 170, 174
    particle size, 106–8
    sulphate content, 102
  ICI catalyst 32–1; 49
  ICI catalyst 32–4; 49–51, 63
  ICI catalyst 35–4; 17, 131, 134–5, 141, 143–5
  ICI catalyst 38–1; 17
  ICI catalyst 38–2; 17
  ICI catalyst 41–3/4; 17, 48
  ICI catalyst 46–1; 18, 65, 78, 81–6, 88
    failure in, 89
    life, 88
  ICI catalyst 51–1; 12
  ICI catalyst 52–1; 12, 110, 114, 117, 153, 171
  ICI catalyst 54–1; 18
  ICI catalyst 54–2; 65, 78, 81, 90
  ICI catalyst 57–1; 18, 65, 78, 81–2, 87–8
    failure in, 89
integral reactors:
  discharging, 178–9
  for catalyst testing, 38–41
iridium, 78
iron:
  catalysts in methanation, 120–1
  crystals in reduction of magnetite, 29
  metallic:
    in ammonia synthesis catalysts, 131–2, 138–40, 143, 172
    in high-temperature shift catalyst reduction, 101, 170
  oxide. *See also* ferric oxide and magnetite
    in desulphurisation, 47, 53
  sulphide, 98, 102–3, 105
    in desulphurisation, 53
isopropanol, 4

K

kalsilite in reforming catalysts, 83
kaolin:
  in reforming catalysts, 80
  supports in methanation catalysts, 121
kieselguhr catalyst, 90
Knudsen diffusion, 116

L

laboratory examination of catalysts, *see* testing
lattice defects, 6

lead poisons in catalysts, 87
least-squares methods, *see* optimisation methods
leptons, 7
lifetime, 5, 20
  ammonia synthesis catalysts, 135, 139
  catalyst testing, 31–2
  desulphurisation catalysts, 48, 50
  high-temperature shift catalysts, 99
  low-temperature shift catalysts, 110, 117
  methanation catalysts, 122
  reforming catalysts, 81–2, 88
lime, *see* calcium oxide
lithium bonds, 132
low-temperature shift, 97, *see also* water-gas shift
  calculations, 217
  catalysts, 109–13, 117, 171–2
    by-passing of, 122
    deactivation of, 111–3
    discharging of, 176
    effect on methanator performance of, 123
    poisoning of, 102, 113, 124–5
    pressure drop, beds of, 206–7
    reactions on, 113
    reduction of, 171, 203
    re-use of, 180
    stabilisation of, 177
  converters, design of, 151–3
  equilibria, 192–5
  reaction, studied by continuous stirred tank reactors, 43

M

magnesia, 9
  ammonia synthesis catalysts, 131, 134
  methanation catalysts, 121
  promoters, 135
  reforming catalysts, 78, 80–2
  stabilisers, 24
  support for reforming catalysts, 79, 82
magnesium aluminate, 133
magnesium hydroxide, 80
magnetite, 29
  absorption of sulphur in, 103
  in ammonia synthesis catalysts, 131–5, 172
  in high-temperature shift catalysts, 98–9, 101, 105, 170
  in redox reaction, 98
  hydrogenation-dehydrogenation activity, 18
manganous oxide, 15
mercaptans:
  hydrogenolysis of, 58
  in steam reforming processes, 84
  thermal decomposition of, 55–6
metals, efficacy of, 10–13
methanation, 4, 18, 97, 117–25, 196
  carbon formation and removal in, 18, 119
  catalysts, 119–21, 172
    poisoning of, 124–5
    pressure drop, beds of, 206–7
    reduction of, 172, 203

    re-use of, 180
    sintering, 124
    stabilisation, 177–8
  effect of voidage differences, 37–8
  equilibria, 118–9, 196
  kinetics, 121–5
  of carbon oxides in cobalt molybdate, 53, 62
methanator purging, 124
methane:
  hydrocarbon feedstock in, 46, 60
  in high-temperature shift conversion, 101
  in primary reformer equilibrium gas composition, 77
  in secondary reformer, 76, 90
  reforming, 64–6, 69, 73–5
    arsenic concentration in, 86
    carbon formation in, 73, 81
    carbon removal in, 75
    catalysts, 81–3, 165, 169, 178
    cracking activities in, 82
    kinetics and mechanism, 90–1
    sulphur concentration in, 84–5
  slip, 86–7, 89
  specific heat of, 128, 131
methane-steam reaction, 72, 87, 90
  equilibrium, 190–1
  calculation of, 94–6, 212–6
methanol, 4
  effect on methanation catalysts, 124
  synthesis in catalyst testing, 43
methyl mercaptan in hydrocarbon feedstocks, 46
micro-meritic analytical technique, 33
molybdenum:
  catalysts, discharging, 176
  loss during cobalt molybdate regeneration, 63
  oxide, 48
monoethanolamine, effect on methanation catalysts of, 124
montecellite in reforming catalysts, 83

N

naphtha:
  desulphurisation, 60, 87
  primary reformer equilibrium gas composition in, 77, 94–6, 212–6
  reforming, 64, 66, 69, 80
    alkali and catalyst acidity in, 82–3
    carbon formation in, 18, 73–4, 81–3, 88
    carbon removal in, 76
    kinetics and mechanism, 91–4
    sulphur concentrations in, 84–5
    -steam reaction, 72
    thiophene concentrations in, 58
natural gas. *See also* methane
  blanketing, 175
  desulphurisation, 46–7
  reformers, 169–70
nickel:
  carbonyl:
    formation, 172
    safety precautions with, 181

SUBJECT INDEX

catalysts, 18
  activity of, 12
  deactivation:
    by halogens, 87
    by sulphur compounds, 46
  disposal, 180
  effect of voidage differences, 38
  in hydrocarbon reforming, 77–9, 82–3, 169
    poisoning, 83–5
  in methanation, 120–1
  in methane-steam reaction, 90–2
  metallic, 172
  safety precautions, 181
  molybdate in desulphurisation, 47
  oxide, 9
    in methanation catalysts, 121–2, 125, 172
    reduction, 122, 169
    in reforming catalysts, 78, 80
nitric oxide in feed gases, 104
nitrogen:
  blanketing, 175, 179
  bonds, 132
  for ammonia synthesis, 90
  for methanator purging, 124
  in low-temperature shift catalyst reduction, 111
  specific heat, 128–9

O

olefins in cobalt molybdate deactivation, 62
optimisation methods for converter calculations, 147–50
ortho silicic acid, 80
osmium in ammonia synthesis, 132
oxidation, 4, 18
  heat released during, 204, 219
  of pellet constituents, 26
  reduction, *see* redox
oxides:
  reducible or unstable, 9
  solid binary, 10–1
oxygen:
  catalyst poisons, 138
  reactions in high-temperature shift reaction, 105

P

palladium, 12, 78
paraffins, steam reforming of, 92–3
pellet:
  size, 22, 100, 106–9
  testing, 44–5
  ultimate tensile strength, 27–8
pelleting catalyst granules, 25–7, 29
  for high-temperature shift conversion, 99, 105
perylene, 14
phenyl mercaptan in heptane desulphurisation, 59–61
phosphorus poisons, 138–9
platinum, 78

poisoning:
  ammonia synthesis catalysts, 138–9, 144, 155
  catalyst testing, 31
  chlorine, 28, 113, 135, 138–9
  high-temperature shift conversion, 104–5
  low-temperature shift conversion, 102–4, 111–3
  methanation catalysts, 124–5
  permanent, 13, 28, 89, 139
    structural collapse, 28
  reforming catalysts, 83–7, 104
  selective, 12
  sulphur, 103, 135
  studied by continuous stirred tank reactors, 43
  temporary, 13, 138
pore:
  diffusion:
    effect of, 20–2, 116, 135
    limitations of, 35, 39, 99
  radius, 6, 7
    ICI catalyst 35–4; 134
  structure in zinc oxide, 49, 63
  volume, 36
    ICI catalyst 52–1; 110
potassium:
  aluminosilicate crystals, 136–7
  carbonate:
    effect on methanation catalysts, 124
    in function of catalysts, 83
    promotors, 135, 138
  chloride in deactivation, 139
  ferrite crystals, 136–7
  hydroxide in functioning of catalysts, 83
  in iron catalysts, 132
  monoxide, 135
  in hydrocarbon reforming catalysts, 78
pre-reduction, 140–1
pressure drop through beds of catalysts, 206–7
  in high-temperature shift conversion, 107–9
primary reforming, 76–7, 79
  catalyst performances in, 87–90
  hydrogen, sulphide formation in, 102
  steam contamination in, 86
programs, *see* computer programs
promoters, 29
  ammonia synthesis catalyst, 126, 132, 134–8, 140
pyrophoric waste treatment, 181

Q

quartz solubility, 80

R

Raschig rings, 99
reactor types, test, 38–43, 145
recirculatory reactors, 41–3, 145
redox reactions, 8, 12, 14
  catalysts for, 97, 115–6
reduction, 9, 29, 168–74. See also pre-reduction

ammonia synthesis catalysts, 133, 139–40
  catalyst strength, 33
  crystal sizes, 134
  heat released during, 203, 218
  high-temperature shift catalysts, 101, 170
  low-temperature shift catalysts, 110–1, 171
  methanation catalysts, 122, 172
  nickel oxide, 122, 169
  oxidation, see redox
  reforming catalysts, 169
regeneration, 174–6
reforming catalysts, see hydrocarbon reforming catalysts
re-oxidation:
  iron, 139, 179
  low-temperature shift catalysts, 114
re-reduction:
  iron, 139, 141
  reforming catalysts, 175
re-use of discharged catalysts, 180
rhodium, 78
ruthenium, 78
  ammonia synthesis catalyst, 132
  methanation catalyst, 121

S

safety precautions, 175–6, 181
sampling of catalysts, see testing
scrubbers in carbon dioxide absorption, 97
secondary reforming, 76–7, 79, 96
  catalyst:
    performances, 90
    reduction, 169
  discharging, 178
  hydrogen sulphide formation in, 102
selectivity, 5
  insulators, 14–5
  metals, 12
  reforming catalysts, 79
  semiconductors, 16–7
semiconductors, efficacy of, 15–7
shift reaction, see water-gas shift reaction
sieving, 163
silica:
  catalysts, 9, 15
  in ammonia synthesis catalysts, 131, 135
  in hydrocarbon reforming catalysts, 78–83
  solubility, 80
  vapour pressure in steam, 81
silver poisoning, 87
single-pass tubular reactors, 41
sintering, 23–5, 28, 139
  ammonia synthesis catalysts, 139, 144
  crystal sizes, 23
  cobalt molybdate regeneration, 63
  nickel reforming catalysts, 87
  methanation catalysts, 124
  thermal, of low-temperature shift catalysts, 111, 114
sodium:
  carbonate, 83
  hydroxide, 83
spacers, 23

low-temperature shift catalysts, 109–10
specific heats, 198–201
spinels in reforming catalysts, 83
stabilisers, 24, 79, 82, 126
stabilisation, 140, 176–9
start-up, 103, 170
steam:
  for methanator purging, 124
  in primary reformer equilibrium gas composition, 77
  reformer calculations, 212–6
strength testing, 44–5
structure, 17, chapter 2, 164
  catalyst testing, 33
  high-temperature shift catalysts, 105–9
sulphinol, effect on methanation catalysts of, 124
sulphur:
  absorption, 49–55
  calculations, 210–1
  catalyst poisons, 84–6, 111–2, 124–5, 138–9, 168–9
  compounds:
    hydrogenolysis of organic, 57–62
    in high-temperature shift conversion, 102–4
    thermal decomposition, 56
  dioxide oxidation, 126
  reactive and non-reactive, 56
  trioxide in ICI catalysts 15–4/5; 102
sulphuric acid in desulphurisation, 47
supports, 15, 17, 24, 110
  alumina, 18, 109
  magnesia, 82
  reforming catalyst, 78–82
  volume of, 22
  zinc oxide, 109
surface, 5–6, 14
  area, 5–7, 21–2
    catalyst testing, 33
    copper, 21, 171
    diffusion limitations, 36
    ICI catalyst 52–1; 110
    loss of, 23, 171
    nickel, 79, 81
    zinc oxide, 49
  reactant transport to, 33–5

T

tableting press, see pelleting
testing, chapter 3
  ammonia synthesis catalysts, 145
tetrahydrothiophene in heptane desulphurisation, 59–61
thermocracking, 78
thermal:
  dissociation of organic sulphur compounds, 54–5
  sintering, see sintering, thermal
thiophenes:
  in heptane desulphurisation, 59–60
  in hydrogenolysis, 58
  in methane reforming catalysts, 84

thermal decomposition of, 55–6
titanium dioxide in ammonia synthesis catalysts, 131
toluene:
  in naphtha feedstock, 88
  reforming, carbon formation, 74
tubular reactors, *see* integral reactors

U

uranium as ammonia synthesis catalyst, 132

V

vacuum extractors, 178–9
vanadium:
  poisoning, 87
  pentoxide, in ammonia synthesis catalysts, 131
Vetrocoke process, 124

W

water:
  catalyst poisons, 138, 140
  -gas shift:
    converter design programs, 152
    reaction, 97–8, 100–1, 192–5
    equilibrium, 94–5, 192–5
    copper catalysts in, 114–5

X

X-ray diffraction, catalyst testing by, 112

Z

zinc:
  carbonate formation in low-temperature shift, 114
  catalysts, disposal of, 180
  oxide, 16, 110
    guard bed, 113
    hydrogenation-dehydrogenation catalyst, 110, 114
    in cadmium evolution, 87
    in desulphurisation, 47–51, 53, 55, 61, 63, 124, 180, 189
    reaction on, 56–7
    support material, 109
  sulphide in desulphurisation, 47, 49, 53, 56
zirconium dioxide in ammonia synthesis, 131

Author Index

Akers, W. W., 90
Andrew, K. F., 74
Annable, D., 142
Apel'baum, L. O., 90–1, 93
Arnold, M. R., 79
Ashmore, P. G., 91
Atwood, K., 79
Auberg, C. H., 61
Baird, R. M., 37
Baugh, H. M., 79
Beattie, J. A., 127
Bodrov, N. M., 90–1, 93
Boreskov, G. K., 42
Bridger, G. W., 65
Brisk, M. L., 43
Camp, D. P., 90
Campbell, J. S., 115–6
Carberry, J. J., 43
Day, R. L., 43
Dent, F. J., 73, 75
Desikan, P., 58
Dirksen, H. A., 79
Dodge, R. L., 127
Dowden, D. A., 17
Earp, F. W., 75
Emmett, P. H., 33
Feitsma, 80
Gadsby, J., 75
Ghoshal, S. R., 60
Gillespie, B. M., 43
Gillespie, C. J., 127
Gilliland, E. R., 74
Goring, G. E., 75
Gromova, U. N., 55
Gulbransen, E. A., 74
Haber, F., 127, 132
Harker, H., 75
Harriott, P., 74
Hasegawa, T., 111
Haywood, D. O., 10
Hedden, K., 75
Hext, G. R., 149
Hill, M. W., 75
Himsworth, F. R., 149
Horn, F., 149
Inoue, T., 84
Isogai, N., 111
Jones, M., 43
Kemball, C., 91

Kiperman, S. L., 142, 145
Kirkpatrick, W. J., 79, 85
Larson, A. T., 127
Livshits, V. D., 142
Luk'yanov, L. I., 142, 145
Lurie, P. C., 79
McKinley, J. B., 57
Malo, R. V., 60
Metcalfe, S. R., 115–6
Moe, J. M., 115
Morita, S., 84
Morozov, N. M., 116, 142
Murphy, D. B., 74
Nielsen, A., 128, 142
Oba, M., 111
Ohtsuka, T., 60
Owens, P. J., 61
Pettyjohn, E. S., 79
Phillipson, J. J., 59
Pichler, H., 84
Pines, H., 14
Pleticka, W. J., 79
Ponnaz, C. Z., 127, 132
Pyzhev, V., 140
Reid, E. E., 55
Richardson, J. T., 48
Riesz, C. H., 79
Roblee, L. H. S., 37
Rudenko, M. G., 55
Schächer, Y., 14
Schäfer, H., 7
Schwartz, C. E., 37
Selwood, P. W., 17
Shapatina, E. N., 142
Shchibrya, G. G., 116
Sidorov, F. P., 142
Smith, J. M., 37
Smyser, H. D., 79
Spendley, W., 149
Stanier, H., 75
Straub, F. G., 80
Tamaru, S., 127, 132
Temkin, M. I., 90–1, 93, 116, 140, 142, 145
Thiele, E. W., 36
Tierney, J. W., 37
Trapnell, B. M. W., 10
Tsaros, C. L., 79
Uchida, H., 111
Vorek, W. E., 60

Warren, J. B., 43
Weisz, P. W., 18
Wheeler, A., 36

Wilson, W. A., 60
Wright, P. G., 91
Wyrwas, W., 65